# EDWIN SELIGMAN'S LECTURES ON PUBLIC FINANCE, 1927–1928

RESEARCH IN THE HISTORY OF ECONOMIC THOUGHT AND METHODOLOGY

Series Editor: Warren J. Samuels

RESEARCH IN THE HISTORY OF ECONOMIC THOUGHT AND METHODOLOGY VOLUME 19-C

# EDWIN SELIGMAN'S LECTURES ON PUBLIC FINANCE, 1927–1928

EDITED BY

**WARREN J. SAMUELS**

*Department of Economics, Michigan State University, East Lansing, MI 48824, USA*

2001

JAI

An Imprint of Elsevier Science

Amsterdam – London – New York – Oxford – Paris – Shannon – Tokyo

ELSEVIER SCIENCE B.V.
Sara Burgerhartstraat 25
P.O. Box 211, 1000 AE Amsterdam, The Netherlands



First edition 2001

Library of Congress Cataloging in Publication Data
A catalog record from the Library of Congress has been applied for.

ISBN: 0-7623-0704-8
ISSN: 0743-4154 (Series)

♾ The paper used in this publication meets the requirements of ANSI/NISO Z39.48-1992 (Permanence of Paper).
Printed in The Netherlands.

# CONTENTS

**BOOK II. Taxation – General**

## BOOK III. Taxation – Special

### Part 1. Direct Taxes: Taxes on Wealth



### Part 2. Indirect Taxes: Taxes on Exchange and Consumption

## Part III. Local Taxation



## PART III. PUBLIC EXPENDITURE



## PART IV. PUBLIC CREDIT

## PART V. THE BUDGET

# LIST OF CONTRIBUTORS

| | |
|---|---|
| *Luca Fiorito* | Via Camollia 73, 53100 Siena, Italy |
| *Marianne Johnson* | Department of Economics, Suffolk University, Boston, MA 02108-2701 |
| *Warren J. Samuels* | Department of Economics, Michigan State University, East Lansing, MI 48824 |

# EDWIN SELIGMAN'S LECTURES ON PUBLIC FINANCE, 1927–1928

Edited by Laura Bruce, Marianne Johnson and Warren J. Samuels

# EDWIN SELIGMAN'S LECTURES ON PUBLIC FINANCE, 1927–1928

Edited by Luca Fiorito, Marianne Johnson and Warren J. Samuels

## I

The notes published below are from Economics 101–102, given in 1927–1928 by Edwin Robert Anderson Seligman (1861–1939) at Columbia University. As with the notes from Seligman's course on the history of economic thought published in Archival Volume 3 (1992), these notes were taken stenographically by P. S. Allen. The notes were found by Luca Fiorito and are published with the permission of their owner: E.R.A. Seligman Papers, Rare Book and Manuscript Library, Columbia University.

As with other comparable archival materials published in these volumes, the notes do not necessarily record what Seligman either said or intended. They must be understood solely for what they are, not a verbatim record of Seligman's lectures but P. S. Allen's notes based thereon. This initial qualification is always relevant to student notes but is particularly important here because of the relatively poor quality of the original notes. The notes reproduced below are *not* the typed original and notice thereof constitutes a second qualification.

These original notes, although also stenographic, are considerably inferior to Allen's notes on the history of economic thought. While they more or less clearly indicate the content being reported on, in some places the notes are disturbingly garbled – often so much so as to make reading difficult and also to make the reader wonder about the native language of the person who took

**Research in the History of Economic Thought and Methodology, Volume 19-C, pages 1–236.**

**ISBN: 0-7623-0704-8**

and/or transcribed the notes. The awkwardness may be due, however, either to casual transcription, with little or no effort to render the text grammatically and syntactically correct, to confusion and/or to failure to understand the material by the original note-taker. Already knowledgeable readers will usually discern meaning from awkward and convoluted wording; others will have to exercise caution. Some, perhaps many, details are confused or otherwise not reliable; however, the main outlines of what seems to be the topics and gravamen of Seligman's lectures come through.

Rather than publish clearly inferior text, we have, with some diffidence, rewritten the notes, using as much as possible of the wording of the original and only the ideas of the original, to make them easier to read. We have tried to improve language, grammar, syntax and punctuation, but not substantive content, the last except with regard to identification of persons and terms. These improvements relate to sentence structure, punctuation, subject-predicate, tense, single-plural agreement (as with "criteria" used in the original instead of "criterion"; also "adapt" instead of "adopt") and other syntactical or grammatical awkwardness. While attempting to make the text more readable, we have tried to both retain the flavor or tone of the language and present what the notes indicate what was in the lectures, or at least heard and taken down. We have attempted to neither seriously revise nor seriously restructure and to neither correct nor make substantive sense of what is ambiguous. (The notes ubiquitously start sentences with "Now", ". . . this is usually deleted.) If the notes are wrong or confused, the error and confusion remains; otherwise, the document we would have prepared would not be the notes. A few mistakes are indicated; some are left as is; many minor typographical and similar errors are corrected without notice, though some are indicated. In other words, we have tried only to clean up and make readable, and to neither fully correct nor make perfect. We have made every effort to avoid both the creation of a different document and the unintentional introduction of interpretations not, or not clearly, present in the original. Of some matters we are too ignorant to affect a revision; in others, we do not want to be overly presumptive in believing that we know what was intended. In places where the awkward language permits more than one interpretation, or we are simply unsure, we have retained the original wording and placed it in square brackets [ ] after our own rendering. Sometimes [sic] is used. Square brackets are also used for editorial insertions. In places where we do not feel confident about editing the language, and because do not want to omit (and designate having done so with an ellipsis), we placed such original text within braces { }. Other slightly ambiguous or mildly awkward wording is left unchanged, though somewhat edited, and unremarked.

Some structural awkwardness has been unchanged. Some lengthy paragraphs, containing multiple topics, have been divided.

We have chosen that route over the alternative of preparing a lengthy detailed summary of our own, in order to provide a readily-available archival record of the (edited) original. The decision to do so was not easy; it was reached only with considerable diffidence. The unedited original is available in the Rare Book and Manuscript collection at Columbia University.

Four further problems. First, the notes have continuously numbered sections. Several of these numbered sections, and possibly some materials, are missing. The "Outline of Lectures", is complete; the missing sections are given in **bold**. Second, the notes are divided into Parts and Books. But the part dealing with public expenditures starts abruptly with designation of neither part nor number; indeed, oddly enough, no Part number seems to be missing. We attribute the foregoing in part to Seligman's omissions of materials and in part to confusion on the part of the person who took the notes and/or the person who transcribed and organized them, who may be the same person. Third, although one part deals explicitly with indirect taxation, no part is so devoted to direct taxation. That topic is treated in section 81. Personal Taxation in Book III. Taxation-Special, of Part II. Public Revenues. This almost certainly reflects the relatively minor roles of income and inheritance taxes at that time. Fourth, the reading assignments are available only for the first half of the course, Economics 101 (Sections 1–44).

## II

Edwin Seligman was both a leading historian of economic thought and a foremost specialist in public finance,[1] active both in academia and the world of affairs (not very many academic economists at the time could say "The Secretary of War was telling me . . ."). Accordingly, early in the course he presents some of the history of the field of public finance, an exercise in both the history of thought and something of the economic, political and social history to which the thought was related. He also examines throughout some of the history of government finance. Seligman compares and contrasts tax and other policies among both the states of the United States and various countries; the course is almost, but not quite, one in comparative public finance.

Different readers will be interested in different materials, statements and arguments. Numerous suggestive insights are provided. The following are a few examples. (For all the quotations in this introduction, we give the original, unedited text [the reader can compare them with our transcriptions]. They typically read better than most of the original notes.) One is an epigram:

> Whenever you find political theories, you find also fiscal theories.

About the U.S., one reads that its

> History during the 19th century was almost blank. We did not have finance, because our problem has been what to do with our surplus. It is a deficit not surplus that creates finance.

– which seems to conflict with the first statement, but which also has a semblance of realism to it.

Seligman follows Adolph Wagner in distilling a sequence of four stages in the evolution of the functions of government. The repressive stage is followed by the preventive, the ameliorative and the constructive stages – and the share of wealth devoted to public, or collective, rather than individual discretion increases, i.e. the growth of public expenditure and of public finance. Apropos of Adolph Wagner's Law of increasing public expenditures, Seligman is quite Wagnerian at several junctures, notably in Part III, Public Expenditures.

Some statements illustrate Seligman's attempt to avoid fallacious or question-begging extreme conceptual formulations. Distinguishing between the Patriarchal and the Contractual theories of the relation between the individual and government, the notes have Seligman treating them as neither adequately positive nor adequately normative theories, but as means of expressing sentiments:

> Just as one magnifies the state, the other minimizes the state, makes it equal with the individual. None of these theories are satisfactory.

The same point is implicit in another rejected distinction:

> Consumption theory assumes that everything government does is bad and reproduction theory says that everything government does is good.

The notes have Seligman contrasting two theories of society – as merely an aggregation of individuals and as having some of the characteristics of a separate entity. These involved conceptual issues which later became very important in much the same way:

> Theory was elaborated by German writers into organismic theory, according to that theory the group as such – does not make any difference what kind of group – the group has a life of its own, soul of its own and feelings of its own they get the idea of super-organism, say church or government. This is absurd because the group is composed of individuals and you cannot say this group is an organism apart from the people that composed this organism. On the other hand you have the phantom public. These writers say you have nothing but a collocation of individuals and there is nothing but individual life. This is just as much fallacy as the other. As soon as the group get together in order, say, to kill elephant if one's idea is different, but it is true that the group is formed of individuals but individuals

> have now a group feeling as well as individual feelings. The individuals in the group have certain idea belonging to the group and at the same time have their own ideas. Starting again with idea, group is neither organism nor a crowd with independent feelings. There is a group existence, in the sense of a group feeling within each individual, therefore, group as such exists.

Seligman is recorded as taking an essentially non-ideological and pragmatic approach to taxation. Rejecting as unsound the extremes of those who think of taxes as good and as bad, he is recorded as saying that

> The mistake lies in the confusion between economy and niggardliness. Reducing taxes is just as bad as over-taxation. The point is that everything necessary is good. If taxes in that sense is desirable, then the question arises, What is the limit?

Seligman's recorded answer to that question seems (it is at least somewhat a matter of interpretation) to emphasize: (1) a concern that the percentage of total taxes to total social income (we would call it GNP or GDP) is not so high as to destroy incentives to produce, (2) a utilitarian approach comparing the advantages and the costs of taxing and spending, and (3) the psychological attitudes underlying the utilitarian comparison and the relation of taxation to incentives.

Interestingly, Seligman distinguishes between the mercantilists of France and England, on the one hand, and those of Germany, on the other. The latter are said to have taken up "primarily the connection between fiscal burdens and general prosperity, or the general welfare" and "were the mercantilists who first elaborated the ideas of public welfare and social welfare".

Seligman also is recorded as saying that "Ricardo was the first economist to attempt a theory of the factory system, based on actual facts; he was the first thinker to deal with matters which are important during the present time". Perhaps Seligman also had in mind Ricardo's theory of value, with its distinction between fixed and circulating capital. The notes, however, record his attention to "the beginnings of that internal struggle between classes – between the landed interest and the moneyed interest", including Ricardo's theory of rent and the controversy over the Corn Laws. As for Ricardo's theories in general, the notes state that "interpretation of Ricardo leads to the conclusion that most of his ideas were hypothetical assumptions built on arguments. His conclusions seem to be rigid and unyielding; that is why Ricardo's theories seem to be absolute. They are thorough so far as they go but they do not go far enough".

Like most people who have worked on the problem, Seligman was beset by the conflicting nuances and definitions of the term "income". Also like most others, he treats it as a positive, or scientific, question, saying that the problem is to "find out what is income". One does not think that Seligman thought of "income" as, in Justice Holmes' phrase, "a brooding omnipresence in the sky". But his language treats it as something existentially (perhaps we should say, ontologically) given, as something to "find" not merely to create with a view to what makes at least some analytical sense, is administratively practicable, and is in some sense fair – though such a view can be intuited from his reported discussion. In this connection, in his discussion of personal exemptions relating to the notion of a minimum of subsistence, the notes report Seligman arguing that the notion of a minimum of subsistence is subjective and that the minimum actually put into the tax law is a "political minimum". Finally, from the standpoint of seventy years later, one can readily intuit the role of various provisions of the tax code in countering problems illustrated in the notes by rival hypothetical cases, e.g. personal deductions with regard to differential ability to pay.

Seligman, in his discussion of whether an appreciation in the value of an asset is to be considered income or, with Irving Fisher, an accretion to capital, treats the matter as a matter of truth. It can be argued, however, that such is not a matter of truth but of social construction, giving effect to some (class) ideology, to some utopian notion of the good society (which may be the same thing), i.e. a matter of policy – even when ensconced in an economic model, definition, or chain of reasoning.

Numerous other interesting and suggestive discussions may be found, depending upon one's interest. For example:

(1) The close and complex historical relationships between land policy, commodification of land, social structure, and public finance – including the process of paying for current expenses by selling off capital assets – as aspects of, in part, the social construction of the economy and the formation and operation of the legal-economic nexus, for example, acquisition from the Indians through confiscation and acquisition by white people through favoritism as the background of subsequent reification of property and other rights (See also item 6).

(2) So far from assuming existing rights, and reaching conclusions of optimality of results tautological therewith – as mainstream economists much later became wont to do – Seligman evaluates arrangements (e.g. forestry ownership) and applies what he considers the social interest, concluding that public and not private ownership is preferable; thereby including legal-economic analysis of property within the field of Public Finance, driven in part by Seligman's broad definition of the field. He also does not hesitate to give his opinions on various aspects of taxation, expenditure, and administrative policy.

(3) In a discussion of government competition with private business intending to raise the level of competitive practice, the conclusion is recorded that "The experience shows us in the case of various instances that instead of pulling up the private competition to the high level of government, the only effect is to pull the government to the low level of private competitor".

(4) An anticipation of a supposedly modern notion, now more often stated in a pejorative way but apparently only descriptively by Seligman, comes in his discussion of the post office letter business: "As to the expediency of state action, the question is, Ought the government run this business? So far as the letters are concerned this has never been questioned. As a general rule, when once government has entered into any business it is very much more difficult to leave that business. People and the community got used to it. The government therefore is almost always in the business 'for keeps'."

(5) The discussion of exemption of government bonds from taxation by other levels of government, using in part the theory of capitalization, makes a great deal of economic sense, but runs up against constitutional provision and interpretation as well as political psychology. The notes are reasonably correct that the situation is due "entirely to political, not economic considerations", though even this, as Seligman acknowledges, is compromised by the economic effects of differential marginal rates.

(6) Correlative to the earlier discussion of individualism and organicism, is a theme manifest in the notes that is more specific than Oliver Wendel Holmes's dictum that taxes are the price paid for civilization. To Seligman, taxes represent the replacement of monarchy with representative democratic government in a world of more or less widely diffused private economic and political power. The practical issues of fiscal politics derive, to no small degree, from efforts to shift tax burdens from one group to another(s) in that world of economic inequality. Aspects of this situation pervade the discussion reported in the notes.

Taxes are the principal way in which government expenditure is financed; taxes may be the price of civilization, more immediately they are the price of government services (whether or not the individual citizen wants any of them). Modern taxes are the replacement of feudal dues, just as modern property is the substitute for feudal property. Seligman thus is reported to have linked the disappearance of the feudal system with the beginnings of economic democracy, including the alienability of land, hence the generation of income from land and selling prices of land, and therefore taxation of property values, either income or capitalized income in the form of market values (see also item 1). Inheritance taxes, for example, are phenomena of the modern system which more or less correlate with/are the modern form of the fees, or fines, which the lord of the land imposed upon the estate of a descendant under the feudal system. If the modern state, at least in principle, taxes at every point at which it can raise revenue, necessary to finance its expenses, the feudal lord imposed a fine, fee, payment, or tax at every point at which he could do likewise. The difference, it may be said, between feudal and modern states in matters of taxation resides in the difference between the status of servant or subject and that of citizen, perhaps precisely in the practices of representative democratic government motivated in 1776 by the claim of "no taxation without representation". (Yet, even in the eventual new United States, various major groups who lacked the right to vote had no (direct) representation: slaves, women, non-property owners. Even today some taxes and charges are voted on only by property owners.)

Seligman is reported to have said that "Montesquieu said that the amount of taxes depend upon the amount of liberty. Free people are willing to pay a great deal more than slaves, and they are more able to do so".

(7) In a discussion of the comparative advantages and disadvantages of private versus government municipal ownership of water and electric light and power operations, one reads:

> You cannot get away from the political system in any way, whether private or public.

(8) In a discussion of then-contemporary patterns of taxation, the notes report Seligman saying that "In certain states we have every survival of the whole system of taxes". This points to both: (1) the hodgepodge nature and incremental mode of development of taxes and tax "systems" and (2) the putative presumptuousness of invoking a survival test of anything in general and of taxes and tax systems in particular. Those who would apply a Darwinian (or unspecified) survival test to institutions often are, it seems, also those who disparage both taxes per se and the existing system of taxes.

(9) Seligman is reported to have attributed the growth of more or less progressive income and inheritance taxes, not to the growth of "socialist" sentiments but to the "democratic movement", a movement apparently left unremarked, or at least not noted. Socialist arguments are seen as going beyond what is called for by the democratic movement: "Of course if you are socialistically inclined, you would say to limit fortunes, but very few go as far as that". Similarly he attributes the growth of public expenditures, in part, to the growth of both democracy and government functions but also, in part, to the growth of capitalism:

> Another point is the expansion of the function of the government. That is due to growth of democracy, also due to growth of capitalism. Also, you remember, the gradual change from repressive to preventive, from preventive to ameliorative, and change from ameliorative to constructive stage.

He concludes that, in spite of efforts to limit the growth of public expenditure, there is and will continue to be

> a continuous increase in expenditures, both absolute and relative, taking place at the present time. So that, in future will still larger importance will be attached to public consumption than private consumption.

(10) In his reported discussion of state-local-federal tax relations, Seligman seems to have adopted and combined two positions, that explaining the contra-Jeffersonian decline of localism and rise of centralism by reference to the growth of a national ecconomy, and that of Adolph Wagner predicting the growth of government and government control. The notes read: "With the growth of national economic interest as over against state and local interest, you are going to have more expenditure and ultimately more control".

(11) Seligman is reported to have said in his lectures on public expenditure, that the scientific study of public finance ("scientific finance") has

> a gap . . . and very largely for the same reason you find a gap in consumption. You see, public expenditure is public consumption; therefore consumption in economic theory ought to be divided into private consumption and public consumption. Now, if you do that you find even in private consumption there has been very little distinction.

This suggests, but by no means proves, that ideology has affected the theoretical handling of public expenditure = public consumption. Interestingly, within a decade, the macroeconomics of John Maynard Keynes treated government spending on new goods and services (in contrast with transfer payments) as a category of spending alongside private consumption, private investment, and net exports (imports) so far as the generation of gross national product, i.e. income, is concerned. Seligman follows this discussion with a strictly utilitarian-pragmetic-empirical, i.e. non-ideological model of government expenditures.

12. Political ideology and system are often treated as if they referred to something ontologically given, part of an independently given and transcendent natural order of things, treating an "is" not only as an "ought" but as a "must". Not so for Seligman. In a discussion of the various modes of the division of tax revenue among levels of government, he is recorded as saying that "There is nothing that naturally belongs to federal or state government".

13. Seligman very occasionally combines a variety of perceptive insights, one or more of which may be cynical or have a potentially cynical twist. For example, he is recorded as saying that "Originally justice was very largely the result of voluntary arbitration, but when the government or king took the matter over it was an expenditure. The king sold justice, and for that reason it was made compulsory. After a long time, recognition of the social function of

justice appears and expenditure increases". The first sentence suggests the nuance of (public) expenditure. The second points to the self-seeking origin of legal institutions. The third recognizes the socialization process.

In comparison with public finance as both a field of practice and a scholarly sub-discipline, the notes indicate a somewhat, perhaps a very, different world from that of today. Although, as is implicit in much of the foregoing, both fundamental issues were raised and complex social processes were discussed, the worlds of praxis and ideas registered in the course were seemingly much simpler, less overtly driven by ideology, more mundane, less analytically esoteric and technical, and more comparative, historical and institutional than exist today, in short, reflective of a seemingly much more innocent period; in fact, a world in which total government spending in the U.S. accounted for less than one-quarter its present share of GNP. Still, while modern terminology and conceptualization – such as option demand, free riders, public goods and externalities – are not used, the issues to which they relate are presented – as in the several principles by which public facilities can be financed. Incidentally, these same discussions indicate something of the large degree of governmental activism in the 19th century in providing infrastructure facilities. Moreover, questions of municipal spending for "public purposes" raised political and constitutional issues, and taxation was always a matter of resolving conflicts of interest and of efforts to shift tax burdens; in such respects, the earlier period was in fact no different from the present.

One aspect of the relatively quaint nature of the topic in 1927–1928 concerns Seligman's recorded statement as to how local taxation in the U.S.

> is the characteristic of the modern life. The immense development of local functions and therefore immense increase of local expenditures for better schools, roads, in civilized life. This is local taxation and proportions of these functions grow broader and wider and brings up the whole problem of local and state finance. How to raise millions of dollars and build bridges, roads and other improvements etc.

Of course, even during the post-World War II period U.S. state and local government spending increased overall at a rate faster than that of the national government.

Further apropos of the twentieth-century growth of government spending, it is widely thought, and more or less correctly so, that prior to World War I the center of attention in the study of public finance involved the shifting and incidence of taxation. With the growth of government spending, progressively greater professional attention was given to such topics as the demand and supply of public spending/public goods and the putative conditions of their efficient provision. Seligman, it should be noted, did publish a leading book on the shifting and incidence of taxation. But these notes from his course suggest that in addition to more or less traditional concerns with marginalist and non-marginalist efficiency, he attended to questions of a structural and distributive kind. These are amply evident in the array of arguments he poses for and against certain policies, arguments which go beyond narrow Paretian considerations and, not to put too fine a point on it, which are seemingly "uninteresting" to present-day theoretical specialists in Public Finance. On the other hand, the inclusion of these arguments could be attributed to Seligman's understanding of the field of Public Finance to be standing at the confluence of Economics and Political Science. Yet, in a sense, that is precisely the point: Taxing and spending policies must, and in fact do, issue forth from more than narrow, technical optimality or efficiency considerations; even when the latter are deployed they typically give effect to implicit normative premises as to whose interests are to count; i.e. distribution and structure govern which efficient result will come to fruition/be adopted. At the very least, therefore, Seligman's evident attention to seemingly mundane matters, matters now largely excluded from technical theoretical Public Finance studies, is evidence of his broader definitions of the field and of public-finance decision making.

Seligman is reported to have provided a very informative discussion of how "real estate" vis-à-vis "personal property" are not given, self-subsistent categories but are matters of legal definition, such that what is real estate in one jurisdiction is not so in another, making tax statistics somewhat non-comparable. Amidst this discussion the notes read: "In other words, the whole uniform feudal system has been modified". The problem is that the idea of a "whole uniform feudal system" is misleading if not wrong. Feudalism varied over space and time, just as modern "property".

The notes show Seligman – sometimes explicitly, more typically implicitly – instructing his audience on several interrelated key points: the actual world of public finance is a mixture of theory and practice, a mixture of statute and administration, a diversity of both general cultures and cultures of public finance, and in every respect comprised of diversity of ends, of means, and of relations between ends and means. There is no single substance or issue on which disagreements could not arise; there were alternative positions on all points. To understand and advise on public finance, it is clearly not enough to rely on either pure economic theory, the canons of property assessment or, inter alia, simplistic ideology. One also needs an understanding both of what amounts to psychology, sociology, politics, history, and so on, and that the levying and administration of a tax involves skills of various types of skilled personnel each with their own professional interests – for example, economists and tax assessors.

One particularly interesting topic is Seligman's interpretation of the status and likely future of property taxation in the U.S. Many of the old issues remain, but not, it would seem, his particular forms of negativism and pessimism.

Much of the foregoing, however, is but an example (or rather set of examples) of how every policy and practice, in juxtaposition to a diverse and complex reality, produces unintended and unexpected results. Which is itself an example of how every policy and practice involves varieties of "good" and "bad" consequences. One of the themes of Seligman's lectures, already amply evident, though only rarely approaching explicit articulation, is that of the *inevitability of problems*. Every tax, every principle of taxation, every method of levying and collecting taxes, indeed everything, has, by virtue of one limitation or conflict or another, from one point of view or another, a problem: "The consequence is that both systems have advantages and disadvantages".

Thus policy analysis and evaluation involves, *inter alia*, both the identification of goodness and badness, or advantages and disadvantages, as such and the respective weights to be assigned to each. All this takes place, moreover, in a world in which people both have an antagonism toward taxes per se, if not also to government per se, and make continued efforts to shift tax burdens to others.

Because of the more or less universal antagonism to taxation, with ubiquitous efforts to shift taxes to others, and with selective adverse psychological reactions to particular taxes, problems and issues multiply. Thus another latent theme arises: There is the domain of public finance theory, of objective analysis and critique. There is also the different world of politics and of political psychology. Tax policy is the object of study by the former, but it is the product of the latter.

A correlative, but subordinate, theme may be that in time enlightened common sense succeeds in making tax policy more sensible – but this is, of course, a matter of selective perception. After all, even the experts often do not agree on important points of theory and of fact.

Where Seligman has his own preferred policy, he is apt to stress the problems consequent to alternative policies; but generally he acknowledges the problems of all policies. For Seligman, public finance has its technical aspects, and they are important, requiring the expertise and cultivated insights of professionals of various types. But for him public finance is ineluctably a subjective matter, ultimately a political matter. Addressing a number of important practical questions, he is recorded as saying that "Every one mentioned is in politics and remains in politics".

In the lectures on public expenditures, Seligman is reported to have several important explicit or implicit themes, including: the fundamental difficulties of financing public facilities, with a variety of modes of raising revenues and a variety of competing rationales, all complicated by efforts both to shift costs to others and to help the disadvantaged; the myth of laissez faire and the reality of government activism; and the practical relevance of political problems and considerations, in addition to narrowly economic ones, even the overriding relevance, if not importance, of the sociology and politics of government finance and expenditure.

On a different aspect, Seligman evidently was fond of instruction by means of classification. He is recorded as saying that "classification is the very first step in scientific process". Although all taxonomies are predicated upon some implicit theoretical schema, they may *seem* less theoretical than modern mathematical formulations. But the latter are themselves often taxonomic in nature or, more likely, give effect to implicit taxonomies and their implicit theoretical schema. Taxonomies, whether explicit or implicit, help organize and define the world, often, if not typically, giving effect to implicit theorizing; after all, the axis on which a taxonomy is located is a function of perception and theory, often perhaps a matter of choice.

It is also the case that this course – a survey course – seems to have been taught for the training less of advanced specialists in the theory of public finance, such as is now the case in graduate courses, and more of the educated citizen including those more or less likely to end up in leadership positions, including public finance administrators.

Seligman was active in practical public finance, at local, state and federal levels and internationally. One respect in which he demonstrated unusual means-ends creativity was in the matter of competitive local under-assessment between counties as each attempted to minimize its citizens' obligations under a county-apportioned state property tax. Seligman proposed, first, that the property tax be a county tax for county, and not state, purposes; and, second, if that was not adopted, to base the division of state income taxes on the relative proportions of assessed valuation, thereby reducing, if not eliminating, motivations for competitive under-assessment.

Finally, we repeat that for all its defects, the edited notes should be of interest to those with an interest in the history of the subdiscipline of public finance. The notes clearly indicate the topics on which Seligman lectured and, albeit less clearly and with some apparent inaccuracy, what he had to say. They tell us much of what was and what was not discussed, as well as something of the level and angle of presentation – though, again, always with the recognition that these are student notes, not the professor's lectures.

We are indebted to Margaret Henderson and Helena Meblack for their help in scanning the document onto the computer. We are also indebted to Bernie

Paris for his encouragement that we limitedly edit the original highly imperfect, difficult-to-read text; to Spencer J. Pack and Andrew S. Skinner for help in locating the quote from Adam Smith; to Harald Hagemann for help in identifying several names; and to the Rare Book and Manuscript Library of Columbia University for permission to publish.

## NOTE

1. Seligman was the author of numerous works in public finance, including:

*The General Property Tax*, New York: Ginn, 1890
*Progressive Taxation in Theory and Practice*, 2nd ed. revised and enlarged, Princeton, NJ: American Economic Association, 1908
*The Income Tax: A Study of the History, Theory and Practice of Income Taxation at Home and Abroad*, 2nd ed. revised and enlarged, New York: Macmillan, 1914
*Essays in Taxation*, 10th ed., New York: Macmillan, 1925
*Studies in Public Finance*, New York: Macmillan, 1925
*On the Shifting and Incidence of Taxation*, 5th ed., New York: Columbia University Press, 1927

Seligman was also the editor-in-chief of
*Encyclopaedia of the Social Sciences*, 15 vols., New York: Macmillan, 1937

# ECONOMICS 101
## Readings in Public Finance
## by Edwin R. A. Seligman

*Secs.*

1–2. Shirras, ch. i; Lutz, ch. i; Bastable, introduction, ch. i; Dalton, ch. i.
3. Bullock, ch. iii; Lutz, ch. ii.
4. Bullock, ch. i; Bastable, ch. ii; Shirras, ch. ii.
5. Adams, introduction; Dalton, chs. ii, iii; Peck, ch. ii.
6–10. Seligman I, ch. i.
11. Seligman I, ch. xiv; Adams, part II, preliminary chapter; Shirras, ch. xiii; Lutz, ch. ix; Bastable, book II, ch. i; Plehn, part II, ch. i; Dalton, ch. iv.
12. Bastable, book II, ch. ii.
13. Lutz, ch. x; Bullock, ch. v; Davies, ch. iii.
14. Adams, part II, book I, ch. i; Hibbard, chs. iv-vi, x, xv.
15. Bastable, book II, ch. v; Dalton, ch. xiv.
16. Shirras, ch. xxx; Bastable, book II, ch. iii.
17. Adams, part II, book i, ch. ii; Madsden, ch. ii.
18. Lutz, ch. xi; Davies, chs. vii, x, xiii.
19–21. Bullock, ch. vi; Smith, chs. i-iii, viii; Barker, ch. xvi.
22. Dixon, chs. ix-xiv; Barker, chs. x-xii.
23. Lutz, ch. xii; Watkins, *passim*; Barker, chs. i-ix, xiii-xv.
24. Bastable, book II, ch. iv.
25. Bullock, ch.,vii; Urdahl, part 1, chs. i, ix, x; part II, chs. vi and viii; Lutz, ch. xiii.
26. Rosewater, chs. i, ii and v.
27. Lutz, ch. xiv; Bastable, book III, ch. i.
28. Adams, part II, book II, ch. i; Dalton, ch. v.
29. Plehn, part II, ch. iv.
30. Seligman 1, pp. 34–37.
31. Kennedy, chs. ii-iv; Seligman I, pp. 38–56.
32. Plehn, ch. v.
33. Cannan, chs. i-iv.,
34. Lutz, ch. xv; Bastable, book III, ch. ii.
35. Weston, ch. iii.
36. Stamp, ch. vi; Bullock, ch. ix.
37. Seligman I, ch. x; Jones, ch. i.
38. Adams, ch. i.

39. Bullock, ch. viii.
40. Seligman II, introduction; Bastable, book III, ch. v; Plehn, part II, ch. x.
41. Seligman II, part I; Dalton, ch. vii; Hobson, ch. iii; Brown, chs. iii, v; Shirras, ch. xviii; Lutz, ch. xvii.
42. Conference I, pp. 424–433; Seligman II, part II, ch. i.
43. Seligman II, part II, chs. ii-viii; Seligman III, ch. ii; Brown, chs. vii-x; Shirras, ch. xix.
44. Seligman III, ch. iii; Stamp I, ch. v; Stamp II, ch. iv; Bullock, ch. viii; Dalton, chs. x-xii.

## REFERENCES

Adams, H. C. (1899). Science of Finance.
Barker, H. (1917). Public Utility Rates.
Bastable, C. F. (1903). Public Finance (3rd ed.).
Brown, H. G. (1924). The Economics of Taxation.
Bullock, C. J. (1924). Selected Readings in Public Finance (3rd ed.).
Cannan, E. (1912). History of Local Rates in England (2nd ed.).
Conference Proceedings of the Conference of the National Tax Association, vols. i–xx, 1907–1927.
Dalton, Hugh (1923). Principles of Public Finance.
Davies, E. (1914). The Collective State in the Making.
Dixon, F. H. (1922). Railroads and Government.
Hibbard, B. H. (1924). A History of the Public Land Policies.
Hobson, J. A. (1919). Taxation in the New State.
Jones, R. (1914). The Nature and First Principles of Taxation.
Kennedy, W. (1913). English Taxation, 1640–1799.
Lutz, H. L. (1924). Public: Finance.
Madsen, A. W. (1916). The State as Manufacturer and Trader.
Peck, H. W. (1925). Taxation and Welfare.
Plehn, C. C. (1926). Introduction to Public Finance (5th ed.).
Rosewater, V. (1898). Special Assessments (2nd ed.).
Seligman I (1925). Essays in Taxation (10th ed).
Seligman II (1927). The Shifting and Incidence of Taxation (5th ed.).
Seligman III (1925). Studies in Public Finance.
Shirras, G. F. (1924). The Science of Public Finance.
Smith, A. D. (1917). The Development of Rates of Postage.
Stamp, J., Sir, I (1921). The Fundamental Principles of Taxation.
Stamp, J., Sir II (1922). Wealth and Taxable Capacity.
Urdahl, T. K. (1898). The Fee System in the United States.
Watkins, G. P. ((1921)). Electrical Rates.

COLUMBIA UNIVERSITY

In the City of New York

ECONOMICS 101–102

# LECTURES

# on

# PUBLIC FINANCE

by

Prof. Edwin R. A. Seligman

(1927–1928)

Stenographic Notes Taken

By P. S. Allen

# PART I

# INTRODUCTION

## 1. Finance and the Science of Finance: Public Finance and Fiscal Science

Economics is the science of business and sometimes the term social economics is used to express an individual in his relation to the group, his wages, how much he acts, etc.; all these depend upon social considerations. In certain respects it is called private economics. Against this private group we have the group called the republic. We call this group a coercive form of association. Just as in the case of a private individual, the government also must look out for its economic life. It must have income; it must live. So we might speak of the public economy over against the private economy. In a general way it might be said that what we have to deal with in this course is the public economy-relation of the citizen to the government, so that sometimes we use the term public finance to express this idea. (The etymology of the word economics: from the Greek words, *ecos*, meaning house, and *nomos*, discipline dealing with the laws of the household.) We can speak of the economics of the government as managing the public wealth and the colonies. Therefore, in the past the term political economy was used to indicate that form of economics which was best suited for free and democratic governments and always included the relation of government to the individual. Nowadays we drop this conception and think of economics from the social point of view.

What we call public finance today means exactly what political economy meant in ancient times. Finance in ancient times meant to pay up, to settle. A very interesting development took place. At the early part of the Middle Ages the word finance in France had two distinct connotations: (1) Finance always meant public finance, public payments. (2) Finance also meant something like cheating. In the Middle Ages it meant unfit abuse. Then came a change for the better. In the first place, finance was no longer exclusively limited to public ideas, and there were a good many financiers, such as the government financiers in France – *Les Haute Finance*. In the second place, after the revolution the old connotation lost its influence and now a man can be a financier without being considered to have any connection with cheating.

In England, after Adam Smith it came to mean primarily private finance and a financier was a banker or money dealer. In Italy, France and Germany the force of early tradition was such that finance always meant public finance. In this country, although the endeavor was made by Prof. H. C. Adams[1] a generation ago to introduce the term science of finance, he was not successful. I propose fiscal science. (*Fiscus* was the government treasure in the Roman Empire). When we speak of fiscal I find that we always mean government or public finance. When we speak about finance we mean Wall Street. What we have to deal with now is Fiscal Science.

## 2. Relation to Economics: Content and Method

What is the relation of Fiscal Science to Economics? It has a double relation. Fiscal science is a part of economics. In another sense it is supplementary to economics. Value, for instance, is a problem of economics; fiscal science also deals with problems of evaluation. Who is paying the taxes? This means you are dealing with value. You have to have theories of wages, rent, profit. Therefore, no one can deal with these unless he has a knowledge of economics. In the other sense it is supplementary. We think of it as primarily being a social discipline, as one man against another man. Whereas economics is primarily private or social and only secondarily public. And so fiscal science is primarily political or public and secondarily private or social. Economics therefore is to be considered from the private point of view and fiscal science from the political point of view. Therefore, they form separate sciences. This is a new development. In olden times, Adam Smith, or J. S. Mill, when writing their books devoted almost the entire book to political economy, devoting only a last chapter or so to fiscal science. But the modern point of view is different.

There are three points of view from which one can investigate the needs of such a study:

(1) It must be done from the point of view of comparative statistics, and your statistics should not be limited; it must be brought into contact with facts so that you can draw your conclusions from information.
(2) You must treat the subject historically. You cannot understand anything unless you understand the original provenance. How do these come to be? What are they?
(3) You must not simply study things as they are, and you must not content yourself only ["at all" in original] as to how it came about. As we all are human beings living in a community, interested in the future progress of pure science, being called the bright lights of actual science, when we are dealing with social science we cannot avert our expectations and hopes; therefore the demand is teleological. Not alone what is, what has been, but also what ought to be. This is not a course in ethics, however; the primary effort must be to bring the problem to light as it is. Moreover, nine of ten economic topics as to what we ought to do are based on what we know.

## 3. **Growth of Public Finance**

At first, there was no state and there was no finance. It was only after states had evolved in order to protect private property and liberty that public needs made their appearance, and it was not long before the public community became identified with the king, etc. Gradually we find security and defense developed and it is only in this later stage that the assumption of the state takes place. In a second sense the great importance of public finance in modern times is due primarily to the industrial revolution. The modern democratic state is the result of the industrial revolution. You can say that public finance is due to three facts:

(1) Growth of wealth in itself.
(2) Change in the functions of government.
(3) Transition in the methods of warfare.

The growth of wealth is a result of the industrial revolution. Agriculture is important – it is worth about 10 billions of dollars in its money value – but during the past few years from out of nothing the automobile industry made its appearance and developed so rapidly that now it is just as important as agriculture.

As for change in the function of government, Wagner[2] said that everything can be explained in these terms: The function of governments is limited to *repression*. War breaks out and we want to end it as soon as possible. In the case of disease or famine, government does its best to terminate it as soon as possible. To these repressive functions we now go one better: we try to *prevent* the occurrence of war or other disasters. All these things are very expensive. There also are several further stages that have to be considered. We are not satisfied with prevention, we try also to improve and to make life better. This stage we will call the *ameliorative* system. Instead of murdering the murderer we put him in a psychopathic institution. This is due to a change in the conception of the form of government. Today, we spend money for good roads and other improvements. All these things cost money. What is true for roads is true for every other thing. The ameliorative stage is more important than the preventive stage. Even this is not enough. In the next stage we create or construct something that does not exist, education as an example. The whole process of education is a chief part of the *constructive* system. In the City of New York alone we spend for education more than the whole of Russia. We spend for this purpose hundreds of millions of dollars and this stage is only in its beginning. So there are four stages and as we realize more and more what the constructive stage of government means, a larger share of wealth will be devoted to the public rather than the individual.

As for the transition in the methods of war, unfortunately war is still with us and will be with us for some time to come. In the time of the Roman Empire, war was the normal thing and peace was the abnormal condition. We have only been able to reverse this situation. By the introduction of gunpowder the methods of war changed, becoming very expensive, and from that time on we have come to know the origins of taxation. Primitive wars did not cost much but the last war was the first time in the history of the world that two things happened: (a) Instead of being a war of governments it became a war of peoples and everybody took part; and (b) that was the first time we ever applied the factory system of production to war. It is estimated that during the last war we spent about 240 billions of dollars. You can realize, therefore, how nearly we were to complete destruction. And, instead of costing a few thousands a day, as it was in ancient times, war now costs millions of dollars a day.

The difficulty in giving statistics regarding the comparative study of different governments or public affairs is their incompatibility. Why are they incompatible? The functions of governments differ from state to state. When you compare one with another you are not comparing the same things. Furthermore, local and general expenses differ, classes of expenditures are not included; in this country we don't include post office expenditures with those of government. Furthermore, the value of money differs. The franc, for instance, now is valued at about 4 cents, whereas a few years ago it was worth about 20 cents. When you go back to the Middle Ages this differs again. Also, most statistics are given in per capita terms. But what interest is per capita data? How can Rockefeller be compared with a street sweeper?

The following figures will give an idea as to the increase of government expenses in different countries:

| | | | | |
|---|---|---|---|---|
| In France: | Chas. V. | 1367 | 8,000,000 | francs |
| | | 1445 | 45,000,000 | |
| | Chas. IX. | 1560 | 84,000,000 | |
| | Louis XIV. | | 100,000,000 | |
| | | 1756 | 253,000,000 | |
| | | 1789 | 475,000,000 | |
| | | 1798 | 750,000,000 | |
| | | 1830 | 2,000,000,000 | |
| | | 1901 | 3,550,000,000 | |
| | | 1913 | 5,000,000,000 | |

After the war was over, instead of 5 billions, we have 57 billions of francs, and now the franc is worth about 4 cents instead of about 20.

| | | | |
|---|---|---|---|
| In Great Britain | 1692 | 4,000,000 | pounds |
| | 1750 | 7,000,000 | |
| | 1790 | 17,000,000 | |
| | 1818 | 58,000,000 | |
| | 1850 | 55,000,000 | |
| | 1860 | 68,000,000 | |
| | 1890 | 90,000,000 | |
| | 1900 | 144,000,000 | |
| | 1910 | 150,000,000 | |
| | 1913 | 197,000,000 | |
| after the world war – | 1920 | 900,000,000 | |

On the other hand, local expenses have increased very much, more extensively since 1900. In 1914 it was 148 millions of pounds, now 350 millions of pounds.

| | | | |
|---|---|---|---|
| In United States | 1791 | 3,000,000 | dollars |
| | 1797 | 6,000,000 | |
| | 1805 | 9,000,000 | |
| | 1825 | 17,000,000 | |
| | 1850 | 54,000,000 | |
| before the civil war – | 1860 | 72,000,000 | |
| after the civil war – | 1872 | 290,000,000 | |
| | 1890 | 358,000,000 | |
| | 1900 | 500,000,000 | |
| | 1910 | 883,000,000 | |
| before the great war – | 1915 | 1,018,000,000 | |
| at the present time – | | 4,000,000,000 | |

| | | | |
|---|---|---|---|
| Taking New York City (local expenditures) | 1789 | 162,000 | dollars |
| | 1800 | 295,000 | |
| | 1814 | 908,000 | |
| | 1860 | 1,681,000 | |
| | 1890 | 7,000,000 | |
| | 1900 | 15,000,000 | |
| | 1915 | 47,000,000 | |
| last year – | 1916 | 186,000,000 | |

| | | | |
|---|---|---|---|
| At the end of century just before consolidation with Brooklyn – | 1900 | 95,000,000 | dollars |
| | 1910 | 133,000,000 | |
| | 1915 | 198,000,000 | |
| | 1920 | 273,000,000 | |
| This year – | 1927 | 516,000,000 | |

## 4. **History and Literature of Fiscal Science**

Because of the lack of literature we will devote more time to this matter.

History of the development of Fiscal Science: The earliest literature perhaps was in India – in old classic India. Most of it is not as interesting as in the Greek writings. The oldest of these is found in the Sacred Books of India, the Code of Menu [Manu],[3] where we simply find statements of facts and is therefore interesting only because it shows the kind of things people believed in at that time. The king's exactions in the form of taxes were looked upon as his wages. The reason a king received wages was because he protects his subjects; people had to pay a certain sum for the benefits they received. So, in the earliest literature we have the idea of benefit of protection. A king who did not give protection to his subjects would be called a thief among the kings. Kautilya[4] was a finance minister to the king of India, and he wrote a great work – Arthasastra [Arthazustra in original] (science of land). This work deals entirely with fiscal questions and part of it has been translated under the name of Kautilya's land policy. More recently a book has been written by Law,[5] an Indian writer. This book contains a series of most amusing comments: Because of the natural weakness of mankind it is quite just to levy taxes in this world as well as fines and other payments. There are also such statements as that liquor shall be sold only in such quantities as not likely to be harmful to others (Paragraph 115). The term begging is used in the sense that the king begs of his subjects.

Chinese Literature: Mr. Chin [?] in his book deals with (1) the economic theory of Confucius and (2) democracy and finance in China. In the pre-Confucian literature you find different schools. About 600 B. C. you have the anarchist school; Chuang Tzu[6] [Hsu Hsuin in original], for instance, was an anarchist. Confucius, who lived between 551 and 477[?], is one of the most outstanding figures of China. He devoted many of his writings to economic questions. The following points are noteworthy: He points out the connection between the burden of government and prosperity. He says, to [word indecipherable: "create"?] offices and establish government for the sake of nourishing the people and to tax them and get revenue from them to support the government. He finds ruling does not increase opportunities. Therefore, he says, at first opportunities must be given to the people to develop all of their resources. The ruler, he says, must enrich the people first and then collect taxes from those who are prosperous. He advocates moderation, certainty, equality of burden and universality. He says the thing to consider is physical welfare, intellectual welfare, then ethical welfare and then the proper conduct of government – the relation of the government to the individual. I want to call your attention to one other passage from Confucius which has led to a great deal of discussion showing that he believed in progressive taxation. He says that both rich and poor must be taxed, such that they both carry the burden of taxation easily. We see therefore that in the Orient, as well as in other parts of the world, human beings are about the same and must obey the same economic rules and conditions.

Later on there was a dispute between Confucius and Huan Quang [?], whose discussions were very important. They turned around the philosophical and social aspects of the government on business and government monopolies. Also a great encyclopedia was published in the 12th century, where problems of expenditures and problems of revenue were discussed very elaborately. Many historical treatises on the history of both philosophical institutions and geographical theories existed in China.

We now come to Greece, where we find the first critical discussions of fiscal problems. Passing over earlier and minor writers I want to call your attention to the leading thinkers. Of these, first and foremost was Xenophon (446–365 B.C.).[7] He wrote a great many works. His *Revenue of Athens* is especially important to us. Written about the middle of the 4th century, it describes the entire system as it then existed and his comments are very interesting. Then, comes Aristotle[8] – the greatest thinker of ["at" in original] all time. He devoted a great deal of time to problems in general, including fiscal problems. The first of his special fiscal treatises is in the work that is called *Oeconomics*. It is really a work in public finance. He discusses three kinds of economics: political economy, public economy and etc. The second book of economics, to a certain degree the most amusing, is a collection of financial expedients utilized by the ruler and by the monopolists in order to make money. [The notes confusingly record discussion of Xenophon's *Oeconomicus* and *Ways and Means to Increase the Revenues of Athens* with Aristotle's *Nichomachean Ethics*.] The philosophical discussions you will find in his works on Ethics and Politics.

One discussion must be mentioned here, because in the Middle Ages it played a great role. He [Aristotle] discusses the problem of fiscal justice. He says there are three kinds of justices: (1) Distributive justice; (2) Commutative justice; (3) Corrective justice. We shall limit ourselves to the first two. By distributive justice he meant the justice that authorizes government in distributing privileges or honors and things of that kind. He says that in dealing with what government does, you are dealing with distributive justice. Commutative justice is used in the sense that justice must be observed regarding the indulgence of human beings in the buying and selling of goods and services and such. With Aristotle the interests of the private person play a great role in commutative justice. He says that in commutative justice you must observe arithmetic proportions but in distributive justice you have to deal with geometric proportions. It means that in the assessment of public burdens you should follow what people can pay. Arithmetic proportion means that poor and rich people should pay the same amount. The same idea goes all through the Middle Ages. Although Aristotle was not a fiscal theorist, he did a great deal with fiscal science and fiscal justice. Plato the idealist, as against a practical scholar, is much more interesting as regards social than fiscal aspects of the problem.

Rome had a very different kind of civilization. The philosophy of Rome was based upon that of the Greeks in fiscal science. Therefore we find very much the same ideas. All the views of Aristotle were present in Rome. According to our knowledge there were very few writers in the field, such as Pliny[9] [Plini in original] and Dio.[10] They contented themselves very largely with elucidating the golden maxims of finance. In fact, we find all through the literature of the Middle Ages continual references of some kind, but unfortunately those maxims were counsels of perfection rather than action. So far as the spirit of Roman law is concerned, there is an underlying spirit in favor of equality and uniformity in public budgets. This is of course in theory, so that in Roman literature there is very little of importance to us.

Arabian Civilization: This was a mighty civilization that existed during the so-called "Dark Ages". Many sciences were kept alive by the Arabs. Ibn Khaldun,[11] the great thinker, wrote a history. He was a philosopher and at the same time a historian. Part of his work is translated into English. His theory was that government should not enter into business to make money. One passage written to General Ibn-el-Hosein [Ibn Khaldun?] states that taxes should be distributed with justice and equality for all, and that no one should be exempted. The ruler should never impose on any one a burden greater than he can bear, for the ruler is the shepherd and guardian of his people. Never take from the people anything except what is superfluous to them, and above all be sure always to spend for their benefit and prosperity and for the improvement of their habits. Kodama [?] in the 9th century wrote a book on taxes, or public revenue, part of which also is translated. Ibn Khordad Bey identifies the geographical, topographical and fiscal conditions of the Empire at that time.

European conditions in the Middle Ages will be discussed under four headings:

(1) Books dealing with administrative and fiscal machinery.
(2) Theoretical problems of justice as discussed by the theorists.
(3) Practical problems as discussed by the layman.
(4) In the course of this period, which had a very short life, are found democratic movements and ideals – essentially modern ideas and discussions.

Discussion of administrative and fiscal machinery: Here we find that there are only three works that are really outstanding. (a) The Domesday Book[12] in England, written in order to facilitate the assessment of taxes and other burdens, includes revenues and appointment, problems, a census, and a statement in detail of what every man owned. (b) The Dialogue of Exchequer, written in 1177, by Niges.[13] This book gave the first accounts of the critical methods of the science, etc. (c) The Book of Census, published by Plini [sic], in which we find a complete record of every piece of land which paid tithes for the support of the church, with comments upon different facts. All these works concern fiscal administration. Caraffa[14] has a work which emphasizes the need of a government treasury and he is very much opposed to government going into business. In England we find almost no lay literature. The only treatise, a very bold one, was written in 1331, entitled *Speculum Regus*.[15] In the next century a work was written by Fortescue [Fortesque in original] (1471),[16] which is extremely important from a political point of view. He discussed special ability and power, subjects and grants.

The economic revolution in Italy and Spain started in the 13th and 14th centuries. In Italy you have the growth of towns, such as Florence and Pisa, indicative of a new economic life with an immense world of commercial interests. We also find there at this time the development of industries. A book was recently published on the history of Florence,[17] the best history of any city ever written. Even at that time cities were so advanced that they could be compared with the American cities of today. There were great banks, trusts, and other institutions, such as we have today. We find the beginnings of the development of political theory. Whenever you find political theories, you find also fiscal theories. Even in the cities of Italy at that time, modern conditions existed. They developed property taxes, graduated taxation and other public revenues which could be duplicated even today. Machiavelli[18] discusses some of these various matters. Guicciardini[19] [Guicciardi in original] in the 15th century discusses the grand councils of France and the pros and cons of taxes; these arguments can be seen in the arguments that go on in Washington. However, democracy was in a very poor condition and after a short time gave way to feudalism. But these 150 years were the most productive in the Middle Ages.

The successors of feudalism were absolutisms, forming absolute monarchies. During the 16th century a certain economic transition took place and by the 17th century economic conditions had engaged such momentum that modern states, as modern England and modern France, appeared. And there came with these developments the appearance of fiscal theories. Then the Reformation and the so-called Humanists appeared. Luther[20] was not by any means a fiscal scientist but he had a great deal to say about fiscal science. He says that the king ought to be to his counselors like a watchman to his horse. Erasmus,[21] with his very enlightened views, was predisposed to have the luxurious class bear the chief burden but in the main the whole movement was directed to strengthening the fiscal powers of the king.

In Germany, M. Osse[22] wrote a great book defining the functions of the king. In 1577, Bodin,[23] a great political philosopher, wrote his great work. He first analyzes the revolution of prices. In one of his books he discusses at great length all the fiscal problems of the day. He says that the king must raise money, operate with wisdom and prudence, and save a part for emergency. He then goes on to discuss six classes of revenues: (1) the public domain; (2) booty; (3) gifts; (4) tributes; (5) state enterprise; and (6) customs and other taxes. He said that when taxes are necessary they are just. He divides taxes into extraordinary and casual. He says that the best taxes are those that are levied upon the things that corrupt people, like luxuries, etc. The remedy for abuse in France he finds in impersonal taxation. Even at the present time this topic occupies great attention. As to loans, he is still in doubt. Bodin dominated all thinking for the following century on the continent and even in England. Frumental Berneau* [?] wrote thick volumes of about 1500 pages very largely against financiers. He said that financiers' hands are so sticky that most of the money running from the people never gets to the king. In Italy Botero,[24] a great writer, wrote a number of books on the subject. In England, Hales,[25] the only

* The name "Frumental Berneau" has not been identified except as follows. Nicolas Froumenteau's *Le Secret des finances de France* (Paris, 2 vols, 1581) is mentioned in Joseph Schumpeter, *History of Economic Analysis*, New York: Oxford University Press, 1954, p. 202n.2, and he may well be the person mentioned by Seligman. That book is an enlarged edition of *Le secret des Thresors de France* attributed to Nicolas Barnard (also Barnaudus), published in three volumes also in 1581, is the same as *Le secret des Thresors de France*, with a third part added. Barnaud was born in 1538 or 1539.

writer who wrote powerful discussions, wrote anonymously in the 15th and 16th centuries. In England as well as in the other countries outside of France this movement began in the 17th century. Most of the discussions centered around protests against ship[ping] money.

The first scientific philosophical mind to deal at large with these questions in England was Thomas Hobbes.[26] He is especially important in finance because he was the founder of the theory of the benefit or exchange value of taxation. He also devotes a great deal of time to problems of expenditures. Hobbes was followed by two men who were economists. The first one who wrote a separate book on economics was Sir Thomas Mun;[27] the second was Sir William Petty,[28] who wrote a separate book on taxation. Petty discusses all the revenues, monopolies, sales, and lotteries. In discussing taxation he closely examines what had developed at that time, the general property tax and the problem of evasion from taxation of personal property. He also has chapters on custom duties, which he calls smuggling objects. Petty advocates the accumulative excise which was brought into England by the Dutch excise upon drinks. This led to a very great discussion in England as to who bears the weight of an excise, the most important question being whether it was the poor man or the rich man. This question was discussed in England for the first time.

John Locke[29] was primarily a philosopher, but he also was very much interested in economic questions. He wrote a great book and dealt also with fiscal questions. He was the originator of the doctrine that all taxes are ultimately borne by the land. Another writer was Davenant.[30] His attention was awakened when England found herself in the war. Davenant found it was pretty hard to finance the war. The end of the 17th century marks the modern point of view as to the connection between social progress and government finance.

In France the matter was much more important, and more development took place; where absolute monarchy failed in England, it succeeded in France. At the time of Henry IV, there was a great deal of discussion on account of Mazarin[31] bringing France into very narrow straits. During the Revolution France was filled with pamphlets and books in which Mazarin and his fiscal policy were attacked – they were called by the French "Mazarinair". This gives a very interesting picture of the ideas and theories of the Reformation. Two very great reformers attempted to stem the tide in France, one was Vauban[32] and the other was Boisguilebert,[33] one of the greatest fiscal reformers of all times.

Boisguilebert undertook, perhaps for the first time, statistical analysis and found the cause of all the troubles to be in the fiscal situation. He wrote a book, called *Details of France*,[34] in which he stated that one-tenth of the population is prosperous, four-tenths reduced to poverty and the rest almost beggars. He made suggestions as to the reform of public burdens and restrictions, in other words, equality of taxation. About the same time the matter was taken up by Marshal de France, Vauban. His attention was called to the economic conditions in France, and he came to the very same conclusion as Boisguilebert. He thought that in view of the multiplicity of taxes the government ought to collect taxes, levying them in kind rather than in money. He wrote a book and called it *Royal Tenth*.[35] His book met with royal disfavor and he was practically exiled. This punishment was able to check all attempts towards fiscal reform in France.

Now we come to Germany. In the 17th century in some respects there were more interesting developments. In Germany we find two groups of writers: (1) the legalist group and (2) the economic group. The legalist group discussed the following topics:

(1) The classification of public revenues.
(2) The fiscal rights of the monarch.
(3) The limits of taxation.
(4) The criteria of a good system.

On the other hand, the economic writers took up primarily the connection between fiscal burdens and general prosperity, or the general welfare. These economic writers are called mercantilists, although they are different from those in France and in England. They were the mercantilists who first elaborated the ideas of public welfare and social welfare.

The works of all of these writers, whether legalistic or economic, were characterized by some rather unpleasant features. There is a great lack of originality, much repetition of platitudes, and quite a number of large books running from 1000 to 2000 pages. These, however, have certain interesting points in common and are in conformity with what was going on in practical life. They upheld the same principles which developed in this country, e.g. relating to the general property tax. They found it the best indication of wealth and the most dependable basis of taxation. They objected to undue taxation of the poor and favored the taxation of the wealthy. They all agreed on the fundamental doctrine that such public demands as existed should be retained. Among the legalist writers we have the three best men: Basold,[36] Maull,[37] and Bornitz.[38] They were very fruitful writers and wrote several worthwhile volumes. Among the economic writers the most important were: Lather,[39] Neumayr[40] [Neumayer in original], Faust [?], Wisemback [?], and Klock.[41] Klock lays down five conditions for equality in taxation:

(1) Taxation must be levied only with absolute necessity.
(2) With equality for all.
(3) In moderation.

(4) Only for public purposes.
(5) Primarily on luxuries.

There are some interesting points there.

That was the first half of the 17th century. In the second half of the 17th century there was a great contest going on between the politicians and the public, the discussion primarily turning on the prerogative of the king. About four or five writers participated in this, such as Seckendorf,[42] Conring[43] [Corning in original], Becker,[44] Schroeder.[45] Large books were written by each of these writers and a recent, very interesting volume appeared in the German literature by a German editor, Zielenziger[46] [Zielengiger in original]. One other book was written by Sommer at the end of 17th century[47] about the old communists. The struggle that existed elsewhere on the Continent had reached Germany, the contest over the excise. That was the literature in the 17th century. Of course, the most important literature in 18th century was that of the cameralists, the Physiocrats and Adam Smith.

In Germany there was a very curious development. Little princelings began to fight with each other and found that they had to depend very much upon their own domain. The matters of the private property of kings and the fiscal administration of boroughs, etc. had to be attended to. Certain offices were created for this and these offices were located in certain rooms. The officers had administrative powers; they were to govern these lands. Thus there grew a kind of administrative force, and when the early professorships were founded they simply tried to explain these administrative perceptions; this science became known as the science of rooms, or chambers – Cameraria. This science soon became more or less identified with another science – police science. Police in those days meant something very different from what it means today. The word police was taken from Aristotle and meant the organizing power of the government. In the Middle Ages the Diet passed ordinances as police ordinances. In the 18th century these became identified with the chambers science. Adam Smith refers in Glasgow to police science, which became identical to cameral science and still later was identified with finance and was used in this sense. German cameral science supplied the detailed material for discussions of fiscal science in its administrative aspects. There are four or five writers, such as Liel [?], Rau,[48] Rohr,[49] but the ablest of them all was Justi,[50] who wrote in the middle of the century.

Justi lays down seven principles of taxation:

(1) A tax ought to be the only appropriate source of wealth.
(2) It must be paid willingly.
(3) It must not interfere with liberty.

(4) It must be equal.
(5) It must be stable and certain.
(6) It should be imposed by smallest number of officials.
(7) The process of collection should be convenient.

His practical suggestions were very much discussed. He thought that the best results would be obtained by imposing taxes on business. He lays down 21 rules for expenditure, and is a little more tolerant. (I will not give you these rules.) Justi is considered to have only one equal and that is Anotria Sonnenfels,[51] who wrote his chief work in 1763. He also makes some interesting additions to the science of police. He gives two principles on taxation: (1) While taxes must be equal, they must not attempt to iron out inequalities of wealth. (2) Taxes must not impair the little capital of the community and taxes must be levied upon social income. He goes on further to distinguish between what he calls depressing and stimulating taxes. He advocates taxes on consumption. Vanderlint,[52] in 18th century Germany, was the first to advocate the use of government revenue to decrease inequalities of wealth. That was a German development.

Now we come to France. As the conditions in France were very much worse, the regent came to the front and invited the public to send in suggestions to remedy conditions. These arrived in such a flood and were so fanciful in character that the bureau in charge was characterized as the bureau of dreams. Among these dreamers were two or three men of common sense. One was Count Boulainvilliers,[53] who published a number of books and had some interesting projects. The second was the famous Abbe St. Pierre,[54] known for his suggestion that a complete revolution in the tax system be made by replacing the old system with taxes on the provinces and on income, what he called the *etuille* [?] tariff. Among the many who wrote books, two or three can be singled out. One was D'Argenson,[55] the author of the famous phrase, "to govern best one must govern least". He classified wealth into: (1) Incipient wealth; (2) Existing wealth; and (3) Persisting wealth.

Next we have to mention the leading Physiocrats: Melon[56] is one who wrote a great deal on economic topics. He is responsible for two rather important fiscal theories that are much discussed even today. He was the first to point to the fiscal distinction between the private individual and the government. He says an ordinary individual must regulate his expenses by his income; government, on the other hand, must regulate its income by its expenses. Melon had a much more adequate conception of credit than his predecessors. He said public debts are really never paid, because it is a debt which is paid from the right hand to the left hand. Montesquieu[57] devoted some chapters to finance and called attention to the relation between taxation and liberty. He said that the more freedom you have, the more taxation you will have; in a nation of slaves it is more difficult to raise taxes. Secondly, Montesquieu was a great advocate of taxes on commodities rather than on wealth. In one passage he seems to favor, for the first time, the principle of graduated or progressive taxation. He devotes a great deal of time to conditions in Greece.

Those are the chief writers, but there remain three or more men of some distinction. We have Dupin,[58] who realized that the situation was bad and made some very interesting suggestions. He discussed in great detail the fact that the land tax ought to be based upon a survey to provide valuation. As far as the towns are concerned, he advocated the imposition of taxes in a lump sum and allowing each town to raise the required sum in any way they liked. Finally, he wrote about adjusting the system of taxation on land, rent and wages. Another writer, Beaumount [?], is famous for his phrases. He says taxes must be like the sails of a boat calculated not to impede or check but rather to lead and assure.

From the middle of 18th century, interest in fiscal matters multiplied, and there were many discussions on both special and general matters. Special projects were due to two individuals: Russell Delatin [?] and Glanieres[59] [Glannier in original]. Both suggested replacement of the existing complicated system of taxes by a simpler method. These suggestions were made but nothing came of them. On the other hand, a movement was undertaken by Quesnay[60] [Kaney in original], the physician to the king. Familiar with the conditions of the peasants, he called attention to the misery of the people and developed some very radical ideas.

We come now to the Physiocratic movement itself. Three great principles of the Physiocratic movement were discussed. Two principles were economic or social, and one was fiscal. The etymology of the word Physiocrat means rule of nature. They said that all of life, physical as well as social, mental or moral, is governed by principles of natural law. The scientific man is to identify these natural laws; the science thus developed was called Physiocracy, and inasmuch as these laws pertained to the economy, they were called the economists, which meant the Physiocrats.

Two fundamental laws were advanced by Physiocrats: (1) Liberty, (2) Prohibition. Their economic analysis concluded that only one productive method of economic life existed; all other activity was variously interpreted in economics. The meaning given to production was the creation of new tangible physical wealth. From that point of view only when labor and capital are applied to land – agriculture – is additional product created. According to this conception every other occupation is considered sterile; agriculture alone produces any profits. They proceeded with an elaborate discussion, working out for the first time a theory of the distribution of wealth and of the influence of changes in price upon the different shares of distribution. Their doctrine was that since every tax has to be paid ultimately out of the surplus and since only land produces the surplus, therefore taxes entirely fall upon land. They argued that there were only two kinds of taxes: direct and indirect taxes. Indirect taxes were uneconomical and inconvenient. It was best to have a single tax directly on land.

Quesnay had a great deal of influence on Count [sic] Mirabeau.[61] Mirabeau became a convert to the doctrines of Quesnay. A number of prominent young men came together and formed the school. There were a great many writers. They included de la Riviére,[62] Dupont de Nemours,[63] Baudeau,[64] and Letrosne.[65] The most important of Quesnay's converts was the greatest economist that lived, Turgot,[66] in some respects even greater than Adam Smith. It is especially interesting for us to remember that when Adam Smith was in Paris he went to meetings of Quesnay and his group. From our point of view of the history of fiscal science, the economic theory of production and distribution, and the theory of incidence and of the effects of taxation upon social burden are after all the chief points of social science. There were some very able antagonists, of course. One of their demands was for free trade, including the exportation of wheat. So far as fiscal science is concerned their great opponent was Graslin,[67] who tried to prove that even a land tax could be shifted under certain conditions; he also advocated progressive taxation.

Beaumont[68] wrote a five-volume study on the fiscal system. He is the only French author quoted by Adam Smith. Another writer was Boncerf,[69] the secretary to Turgot. He was the first man bold enough to question the legitimacy of the entire feudal system. His book was burned publicly and had it not been for Turgot he probably would have been hanged. Voltaire[70] and Rousseau,[71] incidentally, wrote about fiscal and political sciences. There was great movement at this time (1780), and after the downfall of Turgot efforts were made by Necker,[72] a Swiss banker, and also by Calhoun [?], to improve conditions. But everything they did was useless. As you all know, the French Revolution, as well as every other revolution, is caused by fiscal conditions. The writers produced an avalanche of fiscal literature, some of which was of the first class. Some of the writers who contributed to fiscal science were Brissot[73] [Briscot in original], Condorcet,[74] Abbe Sieyès[75] [Sierce in original], Etienne Claviere[76] (1734–1793) [Clabier in original], Roland,[77] Lavioisier,[78] [La Voisier in original] the Younger Mirabeau[79] and Conce [?].

Italy played a great role in the earlier middle ages, but the 18th century was the dark ages of Italy. There are only three men that are worthwhile mentioning: Pascoli[80] (1733), Bandini,[81] about the middle of the century, and

Broggia.[82] But even these writers were more or less rather thin amateurs and were faint reflectors of French ideas.

England. Up to the time of Adam Smith nothing was of significance in England; because England was happy, there was no crisis and the problems that English writers had to deal with were comparatively simple. The only thing of importance was the public debt. Hutcheson[83] was the greatest writer of the time. In 1713 the issues of financial and fiscal science were taken up and led to a great discussion. Over 200 pamphlets were written by a single man. The next great political event was twenty years later, which came about in connection with the famous excise of Walpole.[84] He had very great ambition and introduced a kind of public warehouse. The problem of the excise led to the first important fiscal discussion. At least a dozen works, and many pamphlets, all turned on the question of the incidence and effects of taxation, the tax on tobacco and other things. After that the only project of some writers was to oppose single taxation. Then, associated with Dean Swift,[85] came the question of a separate currency for Ireland. There are only two or three works worth mentioning before Adam Smith.[86] One writer was the great philosopher Hume,[87] whose views on population and wealth are well known. He was also well known on matters of finance. His doctrine of taxation held that taxes, sometimes, instead of being a burden, act as a stimulant when applied properly. Dickson[88] was responsible for the doctrine of the diffusion of taxation. Steuart[89] was the great eclectic writer, responsible for two doctrines: the doctrine justifying the use of the public credit under certain conditions; and, in taxation, the very interesting distinction between proportional and cumulative. By proportional he meant indirect taxation and by cumulative, direct taxation. His general views of economics were those of a conservative nature, and had no influence upon Adam Smith.

We come, therefore, to Adam Smith and the *Wealth of Nations*. Adam Smith is often called the father of economics. It is very absurd to call him such, especially in regard to fiscal doctrines. He is relatively weakest in his writings on fiscal doctrines. In the fifth chapter [sic: Book V, Chapters I and II] of his book he deals with government expenditure and revenues, and even in length it does not occupy many pages. It was his idea that government should not interfere with natural law; accordingly, he has comparatively little to say about the functions of government. His idea was to reduce government functions to the lowest possible limits, even to destroy it. This matter was so much associated with Adam Smith that pretty much every modern discussion in England dates from him. Adam Smith so overshadowed other writers that for the next 30 or 40 years nothing was said about other writers in England. He so dominated the remainder of the 18th century that the few books that followed the *Wealth of Nations* had no significance.

Sketch of fiscal theories in 19th century in England. With the exception of one man – Craig,[90] who wrote a great work in fiscal science – there was nothing to be mentioned until Ricardo,[91] who in some ways had more a acute mind than Adam Smith. He is really the founder of the modern theory of economics. Had Germany remembered her Ricardo, the history of the last 10 years would have been entirely different [undoubtedly referring to the German hyperinflation]. Ricardo lived during a time when England had just emerged from the troubles of the Napoleonic wars. England was almost defeated by Napoleon, and only superhuman efforts prevented it. England then exerted much greater effort than during the last world war; all the problems of war confronted England at that time. In addition to this abnormal situation, there was also the beginnings of that internal struggle between classes – between the landed interest and the moneyed interest. McCulloch[92] wrote a book entitled Principles of Economics [sic: *Principles of Political Economy*, 1825] three-quarters of which was devoted to the fiscal problems – taxation and debt – which primarily interested Ricardo. Ricardo discussed these problems with the expectation of finding general theories of public finance. While he took most of his ideas from Adam Smith and then built his theories of value and distribution, the Ricardian theories were more original. They were not so much under the influence of the Physiocrats as the theories of Adam Smith. Ricardo, however, is very difficult to read. Ricardo was the first economist to attempt a theory of the factory system, based on actual facts; he was the first thinker to deal with matters which are important during the present time. Still, interpretation of Ricardo leads to the conclusion that most of his ideas were hypothetical assumptions built on arguments. His conclusions seem to be rigid and unyielding; that is why Ricardo's theories seem to be absolute. They are thorough so far as they go but they do not go far enough. There are hundreds of matters which are important for us today, however, that Ricardo never touched. Adam Smith and Ricardo between them covered the whole of English thought during this period.

Following this period England dominated the world, leaving no important issues for the English to deal with. John Stuart Mill[93] wrote a book on political economy, but there is nothing new in it. It was mostly on account of the influence that his wife exerted upon Mill that he actually undertook the writing. One other writer is J. McCulloch, who wrote many works. He wrote a book on taxation and makes some very good points, but his was a second-class mind and his work contained nothing new. Jevons[94] never wrote anything worth while about fiscal science. It was Professor Bastable[95] (an Irishman), who studied in Germany, who wrote the first book called Public Finance. It is an excellent book. He wrote in 1890 [sic], just before the fiscal and other reforms; the book is only 37 years old but it is already considered archaic and not up to date. Cannan [Cannon in original][96] wrote a little book on local finance about the end of the century. Even Shaw[97] wrote a book about common sense. That was all the literature in England.

In France after the Revolution there was a reaction. Only three or four writers need be mentioned. The first was an original writer and a genius, who wrote in 1801, *The Principles of Political Economy.* This was Canard,[98] who is the founder of the theory of the diffusion of taxation. This theory was taken up for practical purposes in the middle of the century. Thiers[99] called attention to Canard's theory, saying that no matter where taxes are placed it is distributed among people by the method of diffusion. He won a prize, but his work was an unconscious plagiarism.

The situation in Germany was entirely different. In the first half of the 18th century, Germany was still back in the Middle Ages. There were no developments in literature and science, on account of the thirty-years war. The only activity was the Cameralist movement. But after Adam Smith wrote, a new spirit emerged. After Napoleon had brought them to their knees, Adam Smith's ideas received a good reception in Germany. There was only one man who wrote a book of six volumes that were interesting and worthwhile to read. This man was Professor Rau.[100] Hoffman[101] also wrote a book on taxes. In Germany such conditions developed as a result of the Franco-Prussian war that a host of economic problems made their appearance, such as labor problems, money, railroad and, not least, fiscal problems. As always happens after certain crises there was need for the solution of pressing economic problems. From 1870 on, a situation developed that is comparable only to those of the French Revolution and the Civil War in this country. We, however, shall limit ourselves to fiscal problems.

The youngest of the writers on this topic in Germany was Held.[102] He was a charming personality, and in his 20s he wrote two books about social problems in England. He lost his life saving the life of a young woman. He published a book on the income tax, a little book that lays down the foundation of the modern theories of finance. Wagner,[103] another writer, felt that Rau's six-volume study lacked certain fundamentals and began to study the foundation of the science. He is sometimes considered an exponent of the idea of integrating law and economics. Haeley [?], a student of Wagner's, developed a relativist view of value, property and liberty and, because of his doctrine of relativity, was often called a socialist. He died before he finished his work on public finance. Wagner was the first to show the fundamentals of the socio-political theory of finance. He argued that the function of public revenue is not simply to raise money but also to affect the social theory [to give effect to social theory, or to affect the social order?].

Roscher[104] [Roschsi in original] wrote an exceedingly interesting multi-volume book in which he applied the historical method to public finance. He was the first to describe facts about fiscal science. Another writer, Fenstein [?], who lived in Vienna, wrote a book on finance. His writings are important because he dealt with the comparative and administrative side of finance in all the leading countries of the world: Germany, France, Italy, England, etc. Another writer, G. Cohn,[105] limited himself to the democratic point of view. He wrote a book on the science of finance which has been translated into English. His work is not profound but has a beautiful style. These are men of the first rank. Perhaps the most acute writer was Neumann,[106] who wrote a series of detailed studies. There was an Austrian, Meyer,[107] who wrote the first book on justice in taxation and laid down the principles of graduated taxation. Schwartz [?], who is still living, did not write great books, but was the first to head a scientific journal, the only one of its kind in the world [could this have been G. Schanz (see below)?]. Schoenberg,[108] a very high second-class writer, published a handbook on economics. Sax,[109] who died a few years ago, was the first to attempt to apply the modern doctrines of value and marginal utility.

These and many others worked together in fiscal science. Italians, unable to study fiscal science at home, went to Germany and then utilized what they learned, applying the German ideas. Professor Cossa[110] had the reputation of being a great economist in Europe. His book is translated into English. Ricca[-Salerno][111] wrote the *History of the Theory of Fiscal Doctrines*, going back to the early middle ages in Italy. The most acute mind in fiscal science was Pantaleoni.[112] As a young man he was very remarkable. He spoke four languages. He died only two or three years ago, working out Mussolini's plans. Another man who is still living is Professor Graziani,[113] who wrote what is still the best book in the science of finance.

In America: The history during the 19th century was almost blank. We did not have public finance, because our problem had been what to do with our surplus. It is a deficit, not surplus that creates public finance. It was only after the Civil War that we had such problems. Only one man, D. A. Wells,[114] wrote a pamphlet, entitled "Our Ability to Continue the War". Walker[115] wrote on this topic, though he never understood Wells. Professor Ely[116] started to write on public finance and wrote a very good book. H. C. Adams[117] wrote a book on public debts which made him famous. He also wrote a book on the science of finance. It is a beautiful book, but mostly generalization. Adams devoted most of his time to railroad statistics, and was called to China to do statistical work.

Prof. Daniels[118] wrote a book, and has a reputation of being a stylist. At the end of the century a young man in California, Plehn,[119] published a good book on this subject.

In twentieth century Germany, first came a reaction against all scientific outlook. There were only three men: Heckel,[120] Lotz[121] and Schanz.[122] During the period between 1900 and 1925, work in fiscal science was nonexistent in Germany. After the war, however, there seems to be a renaissance.

In France there is a man of international importance in fiscal science. This is Jèze[123] of Paris, who has translated several of my own books. His work in fiscal science has primarily been about the budget and public debts. He is especially noteworthy as his works are among the best literature in all countries. He is a great protagonist of all modern methods. He is also noted as the editor of the *Review of Fiscal Science and Legislation.* He is a very important man. There is another man to be mentioned who is exceedingly good; this is Allix,[124] who published a two-volume study on the practice of the modern system of public finance.

In England we have an interesting situation. Keynes,[125] Pigou[126] and Scott[127] all wrote during the war. We can also add Hobson,[128] who wrote after the war a volume entitled, *Tax in the Interstate* [sic]. There are three or four men who have come to the front primarily in public finance. Henry Higgs,[129] the oldest, has devoted himself to problems of budgets and management. He writes in a beautiful style; his books have had wide appeal. Hugh Dalton[130] wrote a book which is not very profound but is very stimulating and very suggestive. Most fluent of all the writers in public finance was Stamp.[131] He was for a long time the adviser to the Chancellor of the Exchequer, and ten years ago was the head of internal revenue. He was one of the greatest businessmen of England. Whatever he undertook was successful. He is the promoter of the Dawes plan. He wrote three or four volumes on English public finance. He is not an original thinker but the most remarkable example of the application of fiscal theories to practical affairs from an entirely new modern point of view.

Shirras[132] [Sirrhas in original], Professor of the University of India, has written a book which is excellent and most of the examples are taken from India and everything is fresh.

Italy has produced a very large number of very good first-class writers in public finance. Nitti,[133] who is now hiding in Switzerland, wrote *Science of Finance*, which was soon translated into French. Einaudi[134] has written a large number of books on special topics. He contributed a great many essays during the war to *Corriera della Cerra.* Graziani is a fine thinker, a first-class mind, who has published a five-volume work on the principles of public finance. It is illustrated, interesting and up-to-date.

In Scandinavia the chief economist is Wicksell.[135] His writings on money as well as public finance were really the beginning of modern discussions. He is known in Germany and in England. Pierson[136] [Pearson in original] in Holland is a first class mind. He is President of the Bank and well known in England.

In America we have Professor Bullock[137] of Harvard and Professor T. S. Adams[138] of Yale. Adams has a wonderful gift for administrative work but also has found no time to write books. Professor Hollander,[139] of the Johns Hopkins University, is the best stylist. He writes and speaks most beautifully. He has devoted half or one-third of his time to Public Finance. He is responsible for a great many articles, though he, too, has not made great contributions to public finance. We have no book on the subject in this country. There is one book which is well written. This book is written by Lutz,[140] a student of Bullock's. He is a very able man and we can expect interesting articles from him. Professor Haig [Haynes in original],[141] of Columbia, is perhaps the only man who has an international reputation. Fairchild[142] of Yale is a very good commonsensical man on problems of forest taxation. Professor Brindley,[143] of Iowa, has written some very good works. The most recent works have been done by four or five of my students. Blakey,[144] Professor at Minnesota, is a very good commonsensical writer and a progressive individual. Mr. Peck,[145] Professor at Syracuse, has a book on taxation and finance which is, however, a little exaggerated.

## 5. The General Theory of Public Finance

In the course of history we find a number of different explanations advanced to explain our relations to government and the relation of the government to us. The earliest theory was the Patriarchal theory. Perhaps this theory can be best illustrated from a passage in James Steuart. He says that, "Economy is the art of providing the wants of the family. Such an economy is directed by the head and the steward of the family, but in the state it is directed by servants, but all are children in the state".[146] This theory was applied throughout fiscal science.

Patriarchal theory has two main divisions:

(1) Public household
(2) Private household

Both have heads. The only difference is that the head of state must regard all the subjects as his children. The conclusion is that the government, like the individual, must regulate expenses by its income. This is not, however, in keeping with our times.

Over against the Patriarchal theory is the Contractual theory. This, too, indicates the relation between the individual and government. The individual gives up things to the state in exchange for certain protection. Its best explanation is found in Senior.[147] The exchange theory is one of *quid pro quo*, giving something for something in return. While this is true as regards the group, it is not true for the individual, because the individual does not ask government for support, but is obliged to contribute. Furthermore, even if you could [claim otherwise,] this criterion would not have been a measure of this explanation of government, yet the exchange theory is still found very wise. Just as Hobbes is the originator of the Patriarchal theory, so Rousseau is really the originator of the Contract theory. Just as one magnifies the state, the other minimizes the state, makes it equal with the individual. Neither of these theories are satisfactory.

This was followed by what may be called the Consumption theory. To explain this we have to go back to Adam Smith's idea of production. His idea was, of course, primitive. To produce something, meant for him to create a commodity; his idea is that of absolutely physical production. A merchant calls production the benefit that he creates. Real production by modern economists means anything which brings about surplus over expenses. The modern idea of production includes both physical and intangible goods. Just as there is consumption by the individual, there is also consumption by the community. Just as individual consumption is destructive, so community consumption is destructive, so the best plan of finance is to spend little. The failure to distinguish between different kinds of consumption leads to the idea of different kinds of consumption. There are four kinds, four methods of utilization:

(1) Create a surplus – something new
(2) Eating – maintaining
(3) Wasteful consumption – eating too much
(4) Destructive consumption – use of injurious drugs, etc.

The real distinction is between the four kinds of utilization. We find that this explains public as well as private efforts. The point is, why was the theory of Consumption bound to fail? The logical conclusion cf Consumption theory is that the state is that vast ocean through which every man tries to live at the expense of the public. Against this idea of the relation of the individual to the state, we have the explanation opposed to Consumption theory, the Production theory of finance.

What is the Production theory of Finance? This marks the reaction to Adam Smith by List,[148] the father of tariff protection. His theory is the theory of productive forces. He says that the state – the organized public – exists not so much to produce physical or tangible things in the sense of Adam Smith, but to create conditions to create productive forces. Even the state consumes in order to produce productive forces; government expenditure, or finance, is the best kind of utilization of production. This theory was taken up by other writers, chiefly in Germany, where the state became so powerful and where we find two uses made of it.

First, Dietzel [Ditzel in original],[149] who took up the matter of loans. He said the proper way to maintain the state is by borrowing money. That was his logical conclusion from Production theory, although Wagner and Stamp did not go so far. They called their's the theory of Reproduction, the theory of public finance all of which is opposed to the Consumption theory. Reproduction theory says spend as much as possible and then you will be better off. Consumption theory assumes that everything government does is bad and Reproduction theory says that everything government does is good.

There is still another theory which has been put forward in the last few years. This theory is after the Austrian or Austrian-American School. This is the Marginal Utility theory of finance. The attempt was made to apply this theory to public finance. Since value in exchange in ordinary economic life is to be interpreted in terms of marginal utility, so the relation between individual and government is to be explained in terms of marginal utility. Sax[150] was the first writer to attempt this. Italian, Dutch and Swedish writers began to write about this. This marginal utility theory is nothing but an extension of the second theory, of Contract, a part of the exchange theory.

There are two reasons which show why the application of this fails: (1) It fails to consider the difference between ordinary human beings bartering with each other, in which case of individual against individual there is competition, and the relation between the individual and government, which is not based upon competition. In the case of government, it is a monopoly. Therefore the analogy is false. (2) As between individuals there is always a market price, which depends not only upon competition but upon substitution. In the matter of government, however, there is no such thing as market price.

Sixth theory. If all of the explanations thus far do not explain, what is the proper explanation? The sixth theory is the Positive theory with which every book on public finance ought to deal. How can you explain the relation between an individual and government, primarily in fiscal matters? In order to get this we have to consider the wants of human beings. We all know that when we are dealing with the wants of the individual, some of them he can satisfy himself – like drinking water – unaided, but as soon as there are other persons around him, even though at first wants are isolated, they become socialized wants. You then develop the customs of a community and we learn the acts that meet the approbation of our fellow men. Over against the wants which can be satisfied by means of the actions of an individual alone, there are the wants which can be satisfied only when someone else helps you; we call these wants plural wants. Analyzing those wants further, we find that they belong to two categories. In some cases the wants are different and in others, the same. There are reciprocal wants for exchange. But when wants are the same, as soon as we all want to do something alike, we call them common or mutual wants. We do this through cooperation; therefore we now have a group, with a certain common end. Now, we come to a very difficult point. If, say, each man goes out fishing to his own pond, no other outside help is required, but in case they go out whaling they need the help of others. In this case we cannot be selfish; for our own benefit, we should not be selfish.

Social Theory: We started out with the beginnings of common action and the difference between the ordinary crowd of individuals and the brute. As the result of concerted action the original selfishness of every human being becomes changed in some respects into unselfishness. Now, where we have a group and the emergence of the new ideas of other people, the next question is, what are the activities of the brute as such? The theory was elaborated by German writers into Organismic theory, according to which theory the group as such – it does not make any difference what kind of group – has a life of its own, a soul of its own and feelings of its own; they have the idea of a super-organism, say, a church or government. This is absurd because the group is composed of individuals and you cannot say this group is an organism apart from the people that compose this organism. On the other hand, you have the phantom public. These writers say you have nothing but a collocation of individuals and there is nothing but individual life. This is just as much a fallacy as the other. As soon as the group gets together in order, say, to kill an elephant, if one's idea is different, it is true that the group is formed of individuals but individuals have now a group feeling as well as individual feelings. The individuals in the group have certain ideas belonging to the group and at the same time have their own ideas. Starting again with their idea, the group is neither organism nor a crowd with independent feelings. There is a group existence, in the sense of a group feeling within each individual; therefore, the group as such exists. If the different groups act as a group, what kind of act does it involve? First, to satisfy the common wants of individual, the group satisfies the individuals. There the group is to satisfy common wants. But the group does two other things. In addition to providing you with your wants the group will have relations to other groups and individuals and thus satisfy activities, which may be called quasi-single activities. But the group will do other things: it satisfies your common wants and also satisfies your separate wants as when you belong to a club, you get the privileges of the club and at the same time the club sells cigars, the steward provides you with it. The charge may be little or more, it all depends. Whatever the principle may be, the point is that the group does something else besides satisfying your common wants, it satisfies also your private wants. Therefore we have two kinds of wants: (a) Quasi-single, (b) Quasi-separate. You might say, over against joint activities, the group also indulges in quasi-personal activities.

When we come to the State, over against the above-described groups, we have a group which satisfies the public wants – the political organization. What is the difference between the private and the public group? Other groups are voluntary, state groups are coercive groups. The state differs from private groups in six respects: three characteristics in kind and three in degree.

(1) Fundamentalism. The fundamental things that satisfy your desire for life and protection.
(2) Universalism. The state is the only group which takes everybody in. All other people take part of the people, even the Catholic church does not take everybody in.
(3) Compulsion. In what sense is the state more compulsive, or coercive? When you belong to any group you can dissolve the group and cease to be a member, but you can never get away from the state. Therefore the element of the state which distinguishes it from other groups is its indissolubility based upon the characteristics of fundamentalism and universalism.

There are, however, three other characteristics of the state, and the differences from private groups are matters of degree:

(1) Non-reciprocity. In every other group to which you belong you get something out of it, but in the state there is no such thing as reciprocity, you are born into it.
(2) Indivisibility. In almost every other group what you get out of it you can measure, divide them and measure them, but in the case of the state the advantages are indivisible.
(3) Your relation to the state is immeasurable. There is no way of measuring what you get out of the state.

In all these respects, the public group differs entirely from private groups.

Now, then, we are ready for the conclusion: If the state is a group of that kind, what activity does the state perform and what are the relations for each of us to perform?

The state satisfies public wants which are common wants, as in the case of making war, building roads; but, in addition, like every other group the state has two other kinds of activities: The quasi-single activities, in which the state may buy and sell, run railroads, or make electricity and sell it to you. But the state has not only quasi-single activities, it also has quasi-separate activities, in which the state does something to you to satisfy your separate desires. What shall we call these two classes of activities? These activities are referred to as quasi-public activities.

Now if you will remember these fundamental distinctions, there are two entirely different relations between the individual and government, relations which explain the satisfaction of common wants for which you pay your annual dues. Just as with the corporation, so [with the] public you pay your taxes but over against that the state also has these quasi-public activities, and when you pay for them you don't pay taxes when you buy a railroad ticket and buy water; that is not a tax. There is another conclusion which follows from this analysis, which shows the way out. When you are dealing with public activities you think of the sacrifice of the individual, you think of his ability to pay, you think of the idea of ability. But when you are dealing with quasi-public activities you are thinking of the benefit you get out of it.

The relation of the individual to the state is a double kind and that is the field of Public Science, dealing with both public wants and the satisfactions of the individual. It deals with the wants of the individual in relation to and in connection with both the common wants and the quasi-public or quasi-personal wants. This helps us out of the difficulty. What is the real subject, is it the state or the individual? The subject of fiscal science is not the state in itself, or government; it is the state in relation to the economic and social welfare of the people that compose the public. Those are relations of individuals to government.

We are going to study the nature of these relations, the relations of the individual to government. If a positive explanation is considered, there is one thing to say about the difference between private and fiscal economy. What is the difference between the private economy and the public economy? There are a few differences in kind and a few differences in degree. In the first place, the private economy seeks to secure material goods; the public economy seeks primarily, together with immaterial goods, to create welfare, rather than wealth. Second, the private economy tries to make benefits, but what would the public do with the benefits? The government tries to secure public utilities, it may make benefits but it cannot keep them. The private individual may spend his surplus, the public must spend its surplus, and we have here the fundamental difference. This is in degree. The private economy follows the principle of bargain and sale; now the public may do as the individual does, but only as an incidental and minor part. The public uses the principle of compulsive acquisition and makes you support it, but the private individual cannot make you support him. Government uses expropriation, the private individual cannot expropriate. The private economy looks to the life of the individual. If you borrow money, you have to pay it back. The public economy is unlimited. Its life is perpetual, which makes all the difference as regards finance. In the matter of public debts, it is seen that governments do not make efforts to pay their debts, nor do railroads pay their debts. Sometimes they have perpetual debts. The general theory of public finance is one deserving a great deal of analysis and a great deal of thought.

# PART II

# PUBLIC REVENUES

## 6. **Historical Development of Public Revenues**

There are the following influences: (1) The growing complexity of economic life, which results in the increase of public revenues not only absolutely but also relatively. (2) The growth of the cost of war. (3) The growth of individual wealth, which explains the transition from early revenues, e.g. the public domain, to the more modern system of contribution from individuals. (4) Change in the economic basis of the system from feudalism to modern economic life. In the feudal system land revenues were more important. (5) The transition from absolutism to constitutional government. In the theory of absolutism, the revenues are largely the lucrative prerogative of the king. (6) Growth of economic classes as the various classes assume more importance. With the growth of trade, industry, labor, and associated capital, we find revenues are derived from these sources. (7) Direct taxation and indirect taxation. (8) Growth and the recognition of the community's collective wants, and from the idea of benefit to that of faculty, both of which help explain the transition in the history of public revenues.

We shall now just mention the characteristics of public revenues in different times and in different countries: Most of the taxes originally, especially in the European countries, have been due to war. Other taxes were indirect payments. At that time and even now people are ready to give up their lives instead of their property. As these taxes developed, the natural economy changed into the money economy.

## 7. **The Ancient World: Greece, Rome**

Taking Greece first, during Classical Antiquity, the greatest development was during the time of Pericles in the 5th century. What was the condition there? There were three or four kinds of revenues. There were *tela*[151] [?] – duties that were derived from the public domains. In addition to this there were customs duties and market tolls. The only kind of taxes were on foreign merchants and on slaves. Second in importance were the fines and third in importance were tributes from the subject and allied states. In addition to these they had a peculiar custom which is known *lethargies* [?], which was a combination of honor and obligation. These, however, were only for the extraordinary purposes of war and were called *triarchy* [sic: government by three joint rulers]. In addition to this there were ordinary ones and they were for three purposes: To maintain the chorus, as music played a great role in Greece; to pay the salaries of gymnasts; a survival of the early history of Greece, every year they came together from different cities not as different cities but as members of a clan, and, as the feasting of the crowd required expenses, these were paid by lethargies. This system of the Greeks was entirely different from our present views.

When you come to Rome, you are dealing with 1,000 years of civilization. Taking the great system at the end of the Republic, the revenues were derived mainly from the tributes paid by the provinces. Roman citizens were not required to pay taxes. In later years, however, we find taxes on sales, taxes on law suits, etc. When revenue became more necessary, you find, first, the inheritance tax and the gradual development of a tax on cities, mainly on land; in time, to enforce the tax, a man's property could be confiscated – they even depended upon torture to make people tell where their property was. The first burdens upon the Romans were the local obligations. They were required to do all sorts of things for the government – provide horses for the purpose of helping to build aqueducts, provide clothing for the army, bring grain from Egypt, etc. These were known as burdens and they played a very great role in Rome. They either were connected with property or were personal obligations, or combinations, called mixed. They became so burdensome that freedom from these was considered the best kind of freedom. Therefore, even in Rome conditions were very different from what we have today.

## 8. The Middle Ages: Feudal Income

The essential point and economic contribution of the Middle Ages after the disappearance of the old Roman Empire was the substitution of the Teutonic system of land ownership. The whole Roman commercial and industrial system went back again to the agricultural stage. The basis of the whole of economic life was therefore founded on the manorial system. There was a change from public property to private property, which now became the private property of the feudal lords. The revenue system of the Anglo-Saxon monarchy, a few centuries after the downfall of Rome, can be presented as follows:

(1) Revenues from the public domain.
(2) The benevolences, meaning constitutional aid to the king.
(3) The fines. When a man was killed some fine was taken, one-half would go to the king and the other half to the family of the murdered man.
(4) Right of wreck. If a vessel was wrecked, the remains went to the king. If you find treasure, you could not keep it.
(5) Revenues from the king's farm.
(6) The right to compel to build.
(7) The right of coinage.
(8) Mines of metal belonged to the king.

(9) Markets and tolls belonged to the king.
(10) The protection of stranged [strangers?] (this became very important).
(11) Heriot.[152] When a vassal died, they had to pay certain things to the king.

In England this was summed up as follows: In the towns there were triple obligations: (a) Necessity of fighting wars. (b) Building of bridges. (c) Building of walls. In Ireland and Scotland things were of a more primitive type. During this period we find just the beginnings of taxation. As for protection, they had to build ships to protect themselves against the Danes. On the continent the system was about the same with little difference.

Five centuries later, taking up the 14th and 15th centuries, you come to the development of the feudal system. There you find quite a little change in England at that time; the chief sources of revenues were now as follows:

(1) Revenue from Royal Domain. The king had tenants in the towns and countries. In the towns, the burden on the tenants became so great that they developed a system. They paid a lump sum and began to divide it among themselves in accordance with what they considered to be fair. We have in the second instance the shelter money.
(2) Dowerage.
(3) Relief.
(4) Primer Seisin[153] [Seiser in original].
(5) Wards Chief.
(6) Right of marriage.
(7) Right of escheat. In the case of anyone who died without a will, everything went to the government.
(8) The right of over-lord – *prima noctus, maiden rent.*[154]

The chief fines included:

(1) Fines of the forest.
(2) Fines of justice – court costs today.
(3) Liberties and franchise.
(4) To hold office; another was to leave the office.
(5) Licenses, privileges and concessions.
(6) Fines of protection.
(7) So-called feudal revenues of alienations. These fines were different and a great many fines appeared as counter-fines, concurrent fines, sometimes called queen's gold. These formed very important revenues.

The fourth category was revenue from the royal prerogatives. These were very important in the 14th century. Originally the king was to take a force and support the subjects. But after a while the abuses became so great that it was converted to preemption. When anybody imported, for instance, wine, the king was entitled to take two casks nearest to the mainland. This was the origin of importation taxation. Then we have, secondly, a whole series of payments which were summed up under the name of the casual revenue of the king. Some of these were important, among these were treasury troughs, weights, [indecipherable; possibly "etc".]. These were revenues to the king. There were also the "Royal [indecipherable] right of wreck, and anything salvaged went to the king. If the owner was saved, then he was allowed a part.

The fifth is the Ecclesiastical revenue. The king, being the head of the church, received about one-tenth of all spiritual gifts.

Finally, all these revenues proved to be inadequate and recourse was taken to taxation. Taxation at first was indirect but gradually became more and more direct. What is true in England was usually true on the continent. In addition to the king's taxes, there were also the tax paid to the feudal lord. Feudalism and revenues.

## 9. The Absolute Monarchy: England, the Continent

Feudal revenues gradually were converted into revenues going to the king. When the king became the sole power, naturally certain privileges were given to the king. (a) The sale of offices. (b) Patent of nobility. These were very common and have lasted until the present time. The Catholic church receives these. (c) Still more important was the revenue from granting monopolies. One of the causes of the English revolution was on account of the Stuarts' action in the matter of the sale of these monopolies. (d) On the continent of Europe there was in addition to this, as existed also in England, the right of the community to levy duties upon anything that came into town. Even today in France and Italy these things exist. Also in Scotland was this true. (e) Another was from lotteries. These lotteries lasted until very recently. Most churches, bridges and many other things were built by lotteries in those days.

## 10. The Modern State

We find a very great difference where the king, instead of being lord and master, becomes a paid servant. Summing up the absolute system of monarchical revenue in the order of their relative importance against income:

(1) Public domain.
(2) Royal prerogatives.
(3) Lotteries.
(4) Fines and penalties.

(5) Fees.

In a modern state, the situation is quite the reverse. Public domain plays a slight role, royal prerogatives disappear, fines are of little importance, and fees are less important. The chief thing is taxation.

## 11. **Classification of Public Revenues**

Most of the classifications were made by lawyers and were not made properly. When the scientists began to take up the matter, they had a pretty hard time. Primitive classifications go back to the Middle Ages; other classifications, even the classification by Adam Smith, are not very important. Adams divides the classifications into primary and secondary; Bastable has a good chapter on the history of classification. We will classify revenues into three categories:

(1) Gratuitous payments
(2) Contractual payments
(3) Compulsory payments

Gratuitous payments or revenues are: (a) gifts and donations, (b) subventions and rents, (c) indemnities and tributes.

Contractual payments are the result of a contract expressed or implied between the individual and government and these revenues are prices and charges – traveling on railroads you pay the price, you pay to buy the transportation.

It is in connection with the revenues from compulsory payments that the legal powers became important: The power of judging penalties, the police power – used in the broader sense of governmental powers. The trouble with the police power in economics is that there is no way of distinguishing between the police power and the taxing power. The only advantage is that the courts often have to take refuge in all sorts of fine distinctions. Economically all these payments apart from funds ought to be classified into three heads and those three categories of revenues are called: fees, special assessments, and taxes. What is the difference between each of these? They are all compulsory. The difference between all of these revenues taken together and the contractual revenues is that contractual revenues could be put in business revenues and these revenues are compulsory revenues. In contractual revenues you are dealing with government enterprise but in compulsory revenues you are dealing with government institutions. To buy land you will have to pay so much for an acre, you have also to pay certain fees for land, office, etc. Therefore government acts here in two capacities. To distinguish between a fee, special assessment and tax we have to consider three different points of view: (1) The service that is performed for the individual. (2) The cost of the service to the country. (3) The benefit to the individual. In the case of a fee, this is personal service. In case of a special assessment you could not speak of service to the individual; you have to have the whole area, the assessment for each individual as a member of group. When you talk about a tax it is the service for the community as a whole, as a school house or bridge, etc. From the point of view of cost there is the important distinction in the case of fee; the object is to pay for the cost and not charge you more. In a special assessment they do not bother much about the cost. In the case of tax, there is no rate of cost. It is much more or less. In this case the government tries to get a surplus; in a tax there is always a surplus.

Certain classifications have been discussed, distinctions made between the three points of view: service, cost, and benefit. From the point of view of cost, there were three principles involved: have a deficit, break even, or make a surplus. Government may not get back the cost, and lose money. This is the principle of deficit. The principle of deficit is frequently found in prices – water supply, etc. In the case of fee, the effort almost uniformly is made to recover the cost; for that reason, breaking even is called the fee principle. Of course, in the case of a tax, government gets a surplus, a revenue.

From the three points of view of prices, fees and taxes, in the case of a price there is always a benefit to the individual and in the case of most prices, even in ordinary private life, there is an excess benefit to the individual. In the case of a fee the benefit is still individual but in many cases there is no surplus; you pay about equal value for what you get. In the case of a fee, there might be an excess benefit; there may be a neutral benefit. When we come to a tax, then the benefit becomes converted to a burden, as for instance, when you pay an income tax you will feel the burden.

From all these points of view there are distinctions between prices, fees, special assessments and taxes. In the case of fees and special assessments, there is a little further difference from the point of view of both benefits and costs. In the case of a fee, the benefit is a cost to you personally. In the case of special assessment, there is benefit to you but it is rather insignificant, but when this is for the community the benefit can be larger. There is in this matter a difference between fees and assessment. In a fee it is a more direct service but in the case of assessment it is more to the community.

That being the fundamental question we are now going to take up these various categories. The fundamental difference between taxes on one hand and all other revenues on the other hand. All these others are quasi-public revenues.

# BOOK I

## Quasi-Public Revenues

We are going to deal with quasi-public revenues until Christmas. The first category is price; the second, fees; and the third, special assessments. Because the problems regarding prices are different from the others, we will devote more time to them.

### 12. I. **Prices**

We will take up first the revenue from what might be called public property. If you analyze that, you find that the revenue from public property is divided into four categories which in one way or other are prices. (1) Revenue from property from which no revenue is contemplated. This kind of property we call public institutions, such as hospitals, courts, police, fire stations, schools, etc. (2) Revenue from the public domain, the lands, the other real estate from which a revenue is possible, even probable and is generally actually obtained. (3) Personal property revenue from all sorts of government investments and funds. (4) Revenue from public enterprise or undertakings which government may go into, such as gas business, railroad business, etc. We must make a distinction between public institutions and public enterprises.

In view of its importance in this country, we will begin with the second: Revenue from public domain. The census classification of revenues has been revised to distinguish between general revenues and commercial revenues. Under the heading of commercial revenues are interest from funds, fees and special assessments. Under general revenues are taxes, fines, donations, licenses and permits. Fees generally paid for permits are placed under commercial revenues. The reason is that the census started with municipal statistics then they passed to the government; the classification was suitable for cities but not for the government.

### 13. **Public Domain**

We will discuss this under three heads: (a) History of public domain, (b) Present condition of public domain in this country, and (c) Theories.

The word domain itself comes from Latin form *dominium*. This was synonymous with and referred to the lands that belonged to the community or to the king. Later on it was extended also to other property. In some countries public domain includes more than it does with us. In French the word *domain* has a more general significance. They distinguish between the public domain and the private domain of the state. We never speak of private domain and very often we refer only to land.

Public domain is almost as old as history itself. At one time there was no such thing as private property in lands and therefore you find that in a great many places a large part of the land never became private property. Among the Indians, hunting grounds belonged to the community, also in Switzerland, etc. But when kings developed, they got hold of the land and made use of it. In Greece kings had their public domains. In Rome they played a chief part in the whole constitution. The *agra publica* [public land] was enormous in extent, and everywhere it was given away by kings; by the time of the Empire the *agra publica* [?] was gone. In Greece the chief revenues came from the public domain, such as silver mines, etc. In England the Norman Kings distinguished between ancient demesnes and acquired demesnes. And at one time the king supported himself entirely from the revenues of his lands; when the towns developed, the whole towns belonged to the king. The king then gave his lands away to his favorites. By the end of 17th century a curious custom had developed: one king made presents of the land and his successor resumed it. In the Middle Ages we have both grants and resumptions. A law to further the alienation of all lands was passed at the beginning of the 18th century; and when the king was put upon salary in 1760 the further inalienability of public domain was declared. Since 1760 that is what is left of the public domain in England. In Scotland and Ireland the land was transferred to public wards. In France they had so much more lands, because the principle of inalienability was introduced in 1566 in the ordinance of Mull. During the revolution the property of the church was confiscated; it is called *domain de la nation*. What is true in France is also true in Germany and the result has been that in Germany the condition was more mediaeval, with a larger public domain than in either England or France.

In new countries, like this country and those of South America, the new public domain was gradually turned over to private hands. The situations in North America, South America and Australia are the same historically, but North America started before the others. This explains why the public domain plays a very different role today than it did in ancient Greece and Rome. What are the actual conditions of the public domain?

In England it is known as the Windsor Domains, where the king lives. It also includes some markets, ferries, and forests; the entire income is about 2 or $2\frac{1}{2}$ millions of dollars. Outside of this we have the Duchy of Lancaster which is in 22 different counties and brings in about 60,000 pounds. So in England this is a very insignificant matter. In France the lands and forests are very much greater than in England. They include modern farms and botanical gardens, so that the revenue amounts to a few millions. In the German states it is very much greater. In such German states as Prussia, Bavaria, etc., a very considerable part of the land belongs to the government. One-third of forests belong to communes. Prussia owns salt lakes, forests, etc. There are places in Germany where there are no taxes or income derived from the public domain. On the Continent, to these lands and mines must be added a great many lucrative fishing and hunting grounds. Outside of the European states perhaps the only countries worth mentioning are India and Australia.

In India there has been a controversy as to whether certain revenues from the land should be considered as public domain income or taxes. Is it a return from land or a tax? It is very difficult to say, because in Europe one was developed from the other.

In Australia, very largely under the influence of John Stuart Mill, a system was formed at the time when Australia was developing out of a penal colony. The government decided not to sell the land, but to retain and lease the land. Therefore, public domain leases in all the several Australian states play a great role. They have short leases, long leases, and perpetual leases. In this country, from the initial policy of sale we are going to the policy of lease. The tendency in Australia is to give the land away to get the place populated. Land revenues today derive from leases and sales.

## 14. **The Public Domain of the United States**

The fiscal aspects of our public domain fall under three heads. They are:

(1) Municipal Domain
(2) State Domain
(3) National Domain

1. Public domains originally were local, and when the immigrant settlers founded communities they brought with them the customs and worries of the old world. When the Island of Manhattan was bought from the Indians and settlements were made near the Battery, all the rest belonged to the community and gradually was sold to individuals. In the 18th century, the only municipal revenue, apart from fees and lotteries, came from this public domain. This lasted until the 19th century. Up to that time wells were depended upon for our water supply. When the necessity of a regular water system became evident, a large debt was contracted, and the only way to dispose of it was to sell public domain. We began to sell properties in 1830s. The sales of the public domain were accompanied by the building of water aqueducts and reservoirs. Mr.

Black[155] wrote a book about the early development of the public domain of New York City. He came to the conclusion that it was better for the city to sell the public domain, because government received more income by taxing the commercial enterprises and other institutions that grew then they could ever receive as income from the public domain. The only examples of municipal income are a few parks. In the whole country there are not more than 6,000 acres of municipal public domain. In some other countries, this is more important.

2. State Domain: States in the 18th century ceded all their public domains to the federal government, but certain parcels were reserved and given back to the states for schools and other purposes. This and other lands are still of considerable importance. Outside of public domain reserved for these purposes, only two states have an independent public domain.

In Texas, which came into the union in the middle of the [19th] century, a great deal of land belonged to the state. Consequently it retained the public domain and has had an independent and interesting history, which has been told in various documents. Texas had, up to 1850, 242,000,000 acres and ceded to United States some of that. The history of Texas's public domain may be divided into three or four stages. In the early period they sold large quantities to a few people. In one case an individual got one league. When Texas entered into the Union in 1846, doing what other states did, they sold land at low prices to the settlers. In 1876, when a new constitution was adapted, a little more careful management was decided upon. Especially in recent years oil royalties have been very great. Texas today receives about 4 or 5 millions from its public domain. Outside Texas, the first example of large revenue is Minnesota, which happened to have very valuable forests and iron mines. The constitution of 1857 created a perpetual fund for universities or other purposes. The revenue from the land has been very great; it is growing partly from stumpage and partly from the leases of iron ore. In Minnesota over a million or million and a half dollars is received from the public domain. These are the only states in which any substantial revenue is derived from the public domain.

New York sometimes received some income from salt, but since the middle of last century those sales have been discontinued. In New Jersey the land between the high and low water marks belongs to the state, and since New Jersey is practically annexed to New York City, the revenue from these leases is quite considerable. In the South, phosphate lands in South Carolina and Florida yielded considerable amounts of money. In the same category are oysters in Delaware and Virginia, but the revenues are very slight. Since the discovery of oil in Wyoming the revenues have become rather important. Beyond those examples, there are state parks, which, instead of being a source of revenue, however, are sources of expense. Forests are about the same, with no revenue attached to them. New York comes first in state forests, next comes Pennsylvania, then Idaho and Washington. Municipal and state public domains are, therefore, not very important.

3. Now we come to an account of the federal public domain, which is divided into six sections:

(1) Acquisition of the domain
(2) Disposition
(3) Conservation
(4) Reclamation
(5) Preservation of special features
(6) Fiscal point of view

(1) Acquisition is subdivided into four heads: (a) Gifts; (b) Conquests; (c) Extermination of Indian titles, which we might call confiscation; and (d) Purchase.

Beginning with acquisition, at first there was the treaty of 1783 with Great Britain and later the treaties of 1804 and 1814 and the treaty of Washington in 1842. The first great purchase was that of Louisiana in 1803. Then came the Florida Purchase in 1819; the acquisition of Texas in 1845 and its purchase in 1853; and the purchases of Alaska in 1867 and the Virgin Islands in the last few years. The states ceded 440,000 square miles, as compared to the total area of America of almost two billion acres. By the year 1900 about $1\frac{1}{2}$ billion acres were reserved. Titles had passed to 750 million acres, which left about 600 million acres. In 1925 the unreserved public domain amounted to 186 million acres. How did we dispose of it?

(2) Under the policy of disposition, from the fiscal point of view, there have been 6 stages in the disposition of the public domain:

(a) The purely fiscal stage of disposition from 1784 up to 1801. The government made very large contracts with individuals and companies, which brought about an immense amount of speculation.

(b) The second stage began in 1800, when plots as little as 310 acres were sold, often on credit, at about $2 per acre to be paid in installments. In 1804 it was possible to buy in smaller quantities.

(c) The third stage, begun in 1830, made it a little easier for settlers to acquire land. Since 1820 the government permitted settlers to buy as little as 88 acres; it abolished the credit system and reduced the price to $1.25 an acre. This stage may be called the settlement stage.

(d) The fourth stage began about 1840 and is called the preemption stage. A man could buy in small portions, but not over 300 acres, and he could pay for it after living on it for six months. Then came the great movement which had always been fought tooth and nail by the South, the idea of homestead.

(e) The famous Homestead Act. The idea was to practically give away the land. The land was sold at a minimum cost to home builders. After five years a person could get title to the land. The land was called a homestead and a homestead was immune from any debt of the person. It was under the provisions of this mode of acquisition that a large part of this country was built up. The laws of 1862 were numerous. Under the Timber Act a man could get another 160 acres. In [18 . . . indecipherable], he could get desert land, 640 acres more, and there were other acts that allowing the acquisition of coal or timber land, etc. Under these laws it was possible for a single individual to get 1120 acres.[156] Then came the temptations and speculators. Although many of these acts were eventually repealed, under them several hundred millions of acres were disposed of; by the end of 1890 all the good lands were gone and then began a movement for stopping the wasting of our assets.

(3) We now come to the stage of conservation of our resources. Mr. Pinchot,[157] who was governor of Pennsylvania, went to Europe especially to study this situation, and returned with many new ideas. The second man was Mr. Plunkett[158] [Plunkard in original], an Irishman who got the attention of President Roosevelt, and the President got the whole country behind him. A new policy was introduced, the stage of conservation.

Disposition of Public Domain. After the abuses that developed under the Homestead Act, where land got into the hands of large interests, and especially influenced by recognition of the importance of forestry, a change was brought about in public opinion and a movement formed, first headed by President Roosevelt, but did not lead to any legislation.

In the period of land conservation, some of the more important acts were as follows: In 1909, enlargement of the Homestead Act authorizing various kinds of classifications. President Roosevelt, by executive order, prevented the acquisition of oil lands. A June 1910 law revoked such acts. Another law in 1910 permitted the acquisition of coal and asphalt lands for agricultural purposes but made a distinction between the surface of the land and what was underneath the land. Congress may dispose of the surface but what was underneath the land belongs to the government. This was extended a few years later to oil and gas lands. An important and far-reaching series of acts in 1911 provided for the acquisition of forest reserves by the federal government and the protection of navigable streams. In 1914 a very important act, the gold leasing act, applicable to Alaska, introduced the Australian idea of no more sales, only leases, and prohibited any sale of gold lands. In 1917 similar ideas were applied in the water power act, which reserved from sale all these parts of streams where there were waterfalls for use of electricity, and the mineral leasing act, which has played a great role.

The point to be emphasized is that for the last 10 or 15 years we have entered into an entirely new epoch. Up to 1925 the following amounts have been disposed of. Under the Homestead Act, 220 million acres; Timber Act, 14 million; Desert Land Act, $8\frac{1}{2}$ millions; under the Coal Land Act, a quarter of a billion acres of public domain was in the hands of the Congress in this country. We have a still larger amount given away chiefly for railroads and internal improvements, schools, etc. In 1925 220 millions were given to railroads. You must [add] 134 millions to this, reserved especially for this purpose. What is left in this country is only primarily oil lands, mineral lands, coal lands, and forest lands – land which are today not suitable for cultivation.

(4) Reclamation – Desert land or Swamp Land. Irrigation is an important matter in the history of the world. In this country irrigation was first practiced by the Mormons. But as immigration increased the irrigation scheme proved to be too costly for the individual, and eventually large companies invested millions in irrigation. Difficulties arose that were political and constitutional. The states were unable to carry out this task, and it was recognized, by 1890, that this was a problem which far transcended the power of the individual or a state. In 1902, Mr. Newins[159] succeeded in having his act passed starting reclamation in the desert states. The act provided that revenues from land sales should be put in a revolving fund to be used to finance the construction of reservoirs for storing water which would then be supplied to desert land. Difficulty arose, leading to the amendment of 1914. Reclamation extensions for payments for water rights were extended to 20 years for operation, and maintenance charges imposed per footacre. Divided into irrigation districts, this system became the fiscal agent of the government. The crop value that was irrigated in 1925 amounted to 80 millions of dollars; the total expenditures today amount to 160 millions of dollars. Swamp and overflowing land acts by Mr. Volstead[160] led to the act of 1908, where the drainage taxes were not treated as reclamation. It may be said that we have scarcely begun. The possibility of improvement of swamp land is very great. There is one other feature of the problem which has not yet been taken up in this country. This is with reference to the control of Mississippi flooding. The project of controlling the water and preventing overflow, involves a much bigger problem. This condition is somewhat like that of Egypt, only vastly more important. The problem with the Mississippi is not only a problem of controlling the water, it is also utilizing the flood. We are only at the beginning of our dealing with our public domain. Some day it will be very important to take the land and refertilize it. Properly

treated, this will be a perpetual source of consless [sic] operation, which problem we have not yet been able to visualize.

(5) Preservation and Reservation of Special Features. Here we have primarily the forests. Wood for building purposes has nearly begun to become scarce, due to the depletion of our forests. We began in 1891, with very great hesitation, resulting in Week's act of 1911 and the more recent law of Clark in 1924. Under these laws we now have about 150-odd forest reserves, 175 to 180 millions of acres; we are beginning to realize the advantages of a permanent forest reserve. In addition to this, there are national forests. We have also reserved other areas, not so much for economic reasons as for other reasons, for instance the national parks and national monuments. National parks vary in size and interest, the largest park being the Yellowstone Park, about 2 millions acres. There are a great many parks, all over the West. In addition to forest and national parks, we also have national monuments, under the Interior Departments of Arizona, California and other states. Some of them are under the Department of Agriculture. Grand Canyon is under the Department of Interior.

(6) We now come to the fiscal aspects, where we find some very interesting facts. In the first place, when the public domain was started, the founding fathers, for instance, Jefferson, thought it would be of great importance. As a matter of fact what happened is this. Up to 1882 we received 225 millions of dollars but spent for them $352,000,000. We spent $250,000,000 in buying land from the Indians and for land surveys. We lost $120,000,000. Up to 1900 the revenues have been about $350,000,000, and the cost about $400,000,000. Up to the present the cost runs to about $500,000,000. On the other hand, the one part of the public domain which has been untouched is the forests. Only in the last few years have a few companies begun to introduce improved forestry. If we could manage our forests as well as other countries, we could get from our forests alone about $1,000,000,000. As a matter of fact, our forests, instead of being a resource, are costing us money. We have in the public domain something which will one day become so important that we shall see that Jefferson[161] and Hamilton[162] were right.

What is the theory of the public domain? In discussing the theory of the public domain, you must differentiate between different classes. Take, first, agricultural land. There are three possible dispositions, so far as the fiscal aspect is concerned: (1) Government can cultivate its own domain; (2) it can lease the domain; and (3) it can sell the domain.

The first obviously is not very satisfactory. The government is not a good farmer. We know that from history. Government does not get up early in the morning. The second is also objectionable, because not only does it involve difficult problems connected with short- and long-term leases but it also creates a tenantry, where there is a great deal of embarrassment between individual and government. The third, selling the land, as we have done, create a farming people from whom taxes can be collected. On the other hand, when you deal with other aspects of agricultural land, the conclusions must be modified. Irrigation and drainage leads to the topic of water rights. Our experience shows that private initiative is inadequate. In farming, private ownership is adequate, but public ownership of dams, ditches and canals is preferable to private ownership.

Coming to the forests, it is the social interest that comes primarily in view and seven reasons lead to the conclusion that public and not private ownership is preferable. There are:

(1) Fire protection. Under private ownership, fire protection is inadequate.
(2) Harvesting the timber. Experience shows the great superiority of public control over management.
(3) The betterment of growing crops. If you leave it in private hands, the danger is that the harvest will not be satisfactory.
(4) Water supply. At one time people thought forests created the rain, but now it is known that rain creates the forests. However, the existence of forests covering the land permits the moisture to be stored up gradually.
(5) The utilizing of forest crop.
(6) The improvement of ranch conditions.
(7) For the purpose of recreation chiefly as a play grounds.

When we consider all these factors, it is clear that government forestry is a very much simpler business than government agriculture.

We now come to mines and oil fields. This involves the question of the natural sciences. Turgot wrote a beautiful essay on mines in which he describes all the different ways of dealing with lands. He seems to distinguish between two categories: ordinary minerals and gold, silver, copper, tin and zinc, where we have to consider the nature of monopoly. One ought to distinguish between new mine fields and old mine fields. Here private ownership is preferable. But once a new discovery is made, the question becomes the conservation of the mines, a different matter. We have been using the lease system for a decade or two. It is suggested that these be put on the basis of a patent, or, if this is not possible, apply special taxation. The effort in this country has been to get the large mine magnets on a balance with others. We have not adopted Turgot's suggestion of imposing a certain amount of extra taxation. If you don't do that you get around it with the third plan which we have adapted in the case of oil and coal, e. g., preservation and lease. This is done not for fiscal but for social reasons. Mineral land is thus half way between the forests and agricultural lands.

The final class is parks and monuments, where the argument is in favor of government ownership.

## 15. **Public Property: Funds and Investments**

We distinguish between two kinds of public property, which may be either invested in business or considered by itself. Considered by itself it is called funds and investments. These funds are of many kinds. One kind is in connection with our municipal hospitals, schools, etc. Under the head of budgets are other funds, the accumulation of capital and property in derivative forms. At the present time we have in this country: (1) Public trust funds, chiefly in our states, derived from donations and bequests. The Smithsonian Institution is an example. These funds amount to about one billion dollars. The revenue from these funds come primarily from loans, leases, things of that kind, and they are used for public purposes. (2) We have the so-called public trusts for non-state uses. Some of these are called private trust funds. Hail insurance funds, guarantee of bank deposit funds, cemetery funds, perpetual funds, teachers insurance funds, life insurance funds, maintenance of synagogues funds, etc. (3) We have investment funds. This is not important. In former times the states helped the railways, some of them built railroads. These investment funds amount to about 50 millions. (4) Sinking funds. It is a question which must be discussed more fully, the case of repayment of debt. Taking all public properties together as well as properties that do not bring in any income, they amount to $1\frac{1}{2}$ billions of dollars.

We will not discuss property connected with all kinds of business enterprise, public undertakings over against public institutions. These undertakings are of two chief categories: (a) Administrative undertakings, designed to supply wants such as federal government armor plants and state enterprise, more especially municipal enterprises, such as municipal asphalt business and municipal colleges, and, in some cities, gas and electric enterprise designed as business enterprise. (b) Ordinary business undertakings to furnish the city or individuals with their wants; thus you enter into the whole field of the industry, commerce and finance.

## 16. **Public Industry: Business Enterprise**

In the case of business undertakings or enterprises for service and commodities for public use, it is easily seen that there are two classes: First, you may carry on business for fiscal reason in order to make money, or, second, you may carry on business for social purposes. In the first case we deal with businesses that have generally become monopolies, called fiscal monopolies. As for the social undertakings, the question is, should they be monopolies or not? The danger of competition is very great. When government started railroad monopolies, they did so in order to get rid of the bad competitive practives that had developed among private railroads. Experience shows us in various instances that instead of pulling up private competition to the high level of government, the only effect was to pull the government down to the low level of the private competitor. The argument in favor of monopolies is a strong one. But there are certain exceptions, such as public laundries and public baths. Secondly, where by-products are concerned, there is no reason why there should be a government monopoly. Sometimes government carries on a business not in competition but in order to demonstrate how fine accomplishments can be achieved. The porcelain factories in Germany and France are examples of this.

## 17. **Fiscal Monopolies**

Government does these things purely for financial reasons. There are five of these:

(1) Tobacco
(2) Alcohol
(3) Matches
(4) Salt
(5) Lotteries and miscellaneous

Tobacco monopolies have been known all over Europe for many years. They were introduced in France in the 17th century. Anyone could cultivate tobacco but it had to be under the supervision of the government. It brings a large profit, from 75 to 100 millions of dollars a year. In England they have no tobacco monopoly, they have no tobacco tax either, but no one is allowed to grow tobacco in England. England gets its revenue from customs duties. The other great countries, such as Italy, do not have monopolies; for many years they have leased it out to private companies. They began monopolies in 1884. They have it in Austria, Portugal, Spain and in a great many of the Near Eastern countries, such as Romania, Turkey, and Greece, and also South America. In Japan they had it since the end of the last century, so far as the purchase of tobacco was concerned. In recent years this has been very actively discussed in Germany, Switzerland and Austria. Leasing has never been much in demand in this country, because revenue from taxation of tobacco has been thought large enough.

Alcohol. We call this a fiscal monopoly, as it was not as exclusive as tobacco. Only in South Carolina in this country have we had a state monopoly, and a great deal of revenue was derived from selling whiskey. In Russia, revenues had been immense until 1914, when the system was abolished and prohibition was introduced. Before the 18th amendment we received from the taxation of alcohol more than one billion dollars. This is the amount only of federal revenue; the states also got very important revenue from this source.

Matches: This monopoly is less important. In France they had it in 1872.

Salt. This monopoly is found in Italy, Austria, Greece, Turkey and Switzerland. It has been introduced in Japan and also in a good many colonial governments, such as the Dutch East Indies.

Lotteries. These are a comparatively modern invention. They were introduced in Brussels in 1350. *Tonti*[163] introduced the principle of an insurance policy. This was very popular during the time of Louis XIV. A little later it became a state monopoly all over the world. As for England, I can quote Petty: "Lottery is a tax upon the unfortunate self-conceited fools . . .".[164] In this country lotteries have been very common. It was the chief form of revenue in the 18th century. Many churches and other institutions, even Columbia University, were founded on lotteries. It is only very much later that it was deemed unwise for the state to support any institution out of the income from what people considered gambling. In England state lottery monopolies were abolished in 1820; in France, in 1836. This custom still exists in Italy and in the German states, such as Saxony.

Miscellaneous. We have powder monopolies in France, camphor and amber in Japan, playing cards in Italy and Greece, and opium in India and in a great many Eastern countries. Elsewhere, you have sales of tin, tick wood, coffee, gold, rubber, coconut, etc. More important, however, from the point of view of revenue is the other kind of enterprises.

## 18. **Social Monopolies: Commercial Undertakings**

Here the general feeling is that there ought to be very little done by the government, because of Adam Smith's views that "No characteristic is more incompatible with each other than that of ruler and trader".[165] But our problems are very different today from what they were during the time of Adam Smith. You find a great many businesses and industries tend to private monopoly anyway, and in order to get rid of these evils, the government assumes the business itself. These industries are businesses of fundamental social importance; they have reference to the very basis of our economic and social life not so much with reference to a man getting a living but providing the foundation for getting a living. In a general way, we may say that these businesses deal with the delivery of values and transmission of intelligence. They may be divided into five classes:

I. Transfer of value:
   (1) Coinage
   (2) Issue of paper money
   (3) Banking
   (4) Credit, including insurance
II. Transmission of Intelligence:
   (1) Post office
   (2) Telegraph
   (3) Telephone
III. Transfer of production:
   (1) Market and values
   (2) Docks
   (3) Exchanges
IV. Transportation of value: roads, canals, bridges, railroads, steam and express companies
V. Transmission of utilities and power: water supply, lighting, gas, electric power, steam heat, hot water, irrigation and canals

Taking all these together, consider the problems we have seen with public ownership. The question arises, should the government manage these businesses or should they not? The criteria are the following: The simplicity of the business, the amount of capital invested, and capital management. Also, how effective is the social control of private business? The government may be able to control these properties without owning them.

Before you can answer any question, you better take each problem by itself and apply three rules. Assume that the balance is in favor of government business. Now the problem is a different one. What fiscal principle shall the government follow in managing that business? Here again we have three answers: The government may follow the principle of profits, try to make money out of it; or the government may follow the principle of fees, try to break even, as with the post office; or the government may follow the principle of deficit, even providing a gratuitous service. Government may not want to have ends meet. In Russia, however, government would manage the whole of business. On the other hand, Adam Smith's policy has been *laissez faire*.

## 19. **Coinage, Credit and Insurance**

These businesses in the early period were originally a private matter. Even in the Middle Ages they were a private affair. During the Civil War individuals issued their own money, but ordinarily money has become a function of government. It must be said, however, that in the early periods of the Middle Ages they were operated under the principle of benefits to the kings, called mutation or debasement. When that period of benefits disappeared the government then went on to a fee principle and tried to pay for the expenses of coinage; that is what we call seignorage, used in the sense of one who made money. The French call it brassage. The third stage is to make no charge at all, to permit free or gratuitous coinage. In England this was introduced in the 17th century, in France very much later. In America in 1853 the charge was $\frac{1}{2}$% to 2%; in 1875 it was free. We, however, get large revenues and make large earnings by making coin for other countries. We make a great many medals for institutions, and make profits by selling by-products. We have a very large seignorage, and we get quite a substantial business with our mint. The same thing is true in the case of paper money. The matter of paper money goes back to ancient times, ancient China, France, etc. Laws were passed by England to prohibit the issue of paper money in the American colonies, but during our revolution we issued paper money to such an extent that it was almost a cause of demoralization of the community; this was one of the chief causes that brought about amendments in our constitution. The issue of paper money is the greatest kind of tax.

Banking Business. As far as the fiscal aspects are concerned, you find three views: (1) Run the banks themselves, as in Sweden and Prussia; this is rare. (2) The government may participate in private banks. In New Zealand, the government appoints 4 out of 6 directors. In Argentina, the national mortgage bank is guaranteed as to interest and credential [principal?]. But more common is: (3) Not so much government participation as government control of private banks, what you find in so-called central banks in England, in France and in Germany, also in our system. A certain portion of the profits, say 10%, goes to the government. Outside of the Federal Reserve Banks we have the Farm Loan system under the act of 1916. We get revenue not so much from these branches, but, in the case of transmitting credit to the banks whose capital is small, say 5 millions of dollars, the government gets 50% of its income. Taking everything together, our revenues from the banking system are considerable. In 1926 it was about $20,000,000.

A common form of participation in the transfer-type of business is in insurance. Many countries, such as Germany and England, carry on insurance against unemployment. New Zealand and Italy have life insurance. In this country the greatest example is our war-risk insurance. Some states have systems of insurance all of which bring in revenues. Wisconsin and New York have accident insurance, South Dakota has hail insurance. The object, however, is social rather than fiscal. The much more important transfer-type of business is the Post Office.

## 20. **Post Office, Telegraph and Telephone**

We will discuss this problem under four heads: (1) History; (2) Kinds of Business; (3) The expediency of government action, should the state run those businesses? (4) What should be the principle of charging customers?

History. The post office is a modern institution. It is not found in Greece or Rome. The development of the Hanseatic towns in the 14th century led to the building of post offices. With the development of universities in the 13th and 14th centuries, students came from different countries and each student had need of communication with his homeland. Out of this grew the modern post office system. Government made a monopoly of it, beginning in France in the 17th century and about the same time in England, where in the 18th century coaches were introduced and only later were modern methods applied. Early charges were enormous; for example, small packages from London to Dublin cost about 50 pounds. The charges were paid on delivery. In this country we had a colonial post, later on we had an intercolonial business. Benjamin Franklin was interested in this. Our charges were very high and it took a very long time, about a week, to get a letter from New York to Boston. Of course in those days there were very little in common among the people, so development was very slow.

The great significance of the postal system, fiscal and otherwise, is due to the change in 1840. Rowland Hill, a young man, who had become engaged to a young lady, tried to communicate more often but on account of this being an expensive proposition, he worked out a scheme and proposed the following plan: (a) There should be only one charge. (b) One uniform charge to any place. (c) Charges should be prepaid. (d) Postage stamps should be used. This was referred to as the most "idiotic scheme", and after a great hesitation the method was tried. The revenue at first fell off, but the number of letters carried increased more than ever and the system became very popular. In 1845 we adopted uniformity of system, in 1847 we adapted prepaid stamps, and in 1851 the low rate of postage, 2 cents and 3 cents, then finally 2 cents and 1 cent. By 1851 these four principles had been adopted. That is the basis of the whole modern post office. Then began the international post. A young German official first suggested it in the early 1860s; however, it was not until 1874 that the international convention of Bern laid down the foundations of the international post. However, we still do not have a universal postage stamp.

We have been so far speaking of the business of transmission of letters. We also have the Passenger Post business. Parcel Post is another; postal money order is still another, which started in this country in 1846. The C. O. D business is again due to German initiative. The savings bank business, postal telegraph, and postal telephone – altogether eight different kinds of business have come under the post office business. Rural delivery started in 1896. Special delivery started in 1885.

As to the expediency of state action, the question is, ought the government run this business? So far as letters are concerned this has never been questioned. As a general rule, when once government has entered into any business it is very much more difficult to leave that business. People and the community get used to it. The government therefore is almost always in the business "for keeps". The main interest is that the government ought to run the letters, and apply three principles: (a) everybody is interested; (b) no capital outlay; (c) there is very small danger of official sleepiness. Even then, however, if one of the postmasters general had to run a private business he could run it only on a one-third basis. The argument is that the government ought to run the post office. If the government ought to run the post office, how should it be run? Shall the government run the business on the principle of benefits, fees, or gratuitous service? Most of the European countries try to make money out of it.

In this country we have always taken the opposite view. We have adopted the principle of fees; we try to make both ends meet. When we have a surplus, we devote it to increasing facilities, or reducing rates. Last year we had a 40 million dollar deficit. The Secretary of State and Postmaster General are in such a relation that only when there is a deficit, he has to pay and if there is any surplus it is turned over to the Secretary of State. Gratuitous service is not practicable because of the fact that one man receives, say, one letter in a month, another person receives one thousand letters, which will not be just under this principle. We charge on the letter business a pretty high rate inasmuch as a letter is more important and, therefore, we make immense profits on first class mail. But these are more than eaten up by reverse expenses on second, third, and fourth class mail. It is wrong to suppose that this difference is entirely based on value. If you register and use special delivery, you pay more; the result is that probably it will take a little longer to break even, we again make benefits and increase facilities, etc. Our postal budget will be about one billion dollars. There is no reason why the post office system should be treated any differently from any other function of government.

Telegraph. History: It was invented in this country in 1884 by Mr. Morse. He succeeded in starting a line between Baltimore and Washington. Its business for the first [week] was 73 cents. The Postmaster General said government operation was out of question because of the expense. The second argument was more convincing: at that time there was a reaction against all government enterprises. The telegraph has existed in this country as a private enterprise. The same is true in England. In 1868–69, when the situation was very much like that of post office, a great movement began for cheaper telegraphy, which resulted in the government taking over the telegraph. Low rates were introduced, 6d. all over the country. The consequence was that, as in the post office, the revenue fell off, although the number of telegrams increased. In the case of post office the slack was taken up but in the case of telegrams the slack was not. The telegraph in England showed a deficit. As a result of the English movement, agitation in favor of government operation began in this country, but it was finally overcome. Since then we have had no discussion of the issue, in-as-much as the telegraph companies have been very careful to give priority to government business and also to satisfy their other customers. The only complaint is that it is a little too expensive. Everywhere else in the world, the telegraph is part of the post office system; by utilizing the mail there is a very great advantage in the European system. In the way of charges we are very much behind and not ahead. In the telegraph they do not try to make money as they do in the post office in the European countries.

Telephone. Graham Bell in 1874–76 perfected it, and Edison put the final touch upon it. It has a wonderful system of management and it was possible to combine the telegraph and telephone to be better serviceable. In this country, therefore, there has never been a demand for government enterprise. In Europe the situation was different and the telephone either began as part of post office, as in France and England, or was taken over by government. Government took over all the telephones in 1910. There are conflicting arguments. On the one hand, rates are much lower abroad; they haven't abolished the zone system. People use the telephone much more. With us, on the other hand, although rates are very much higher yet the service is so inestimably superior in this country, and as regards the efficiency of management there is no comparison at all; the choice therefore is not great. In the case of the telegraph the argument is in favor of government business. In the case of the telephone the argument is much more difficult. We approve of the expenses on the principle that there is economy in high prices just as there is economy in high wages. The question is entirely one of relative advantage. Under these circumstances when there is so much prosperity, the undertaking will not go to government, but as soon as there should be something of difficulty or bad times the government might be asked to take the enterprise itself. There are a great many problems affecting the fiscal aspect but these businesses are after all private businesses.

## 21. Markets, Docks and Harbors

Markets: In countries like the United States the only example we have of markets is municipal markets, ownership and management by the municipality. We get revenue from our markets but the management is in the hands of private individuals.

Docks and Wharfs: We see both in state and municipal finance. California receives several million a year, whereas in most parts of this country docks belong to the city. Other state ownerships are in Massachusetts, Connecticut and Rhode Island. On the other hand, in some cities we have very large docks; the chief example is in this city. About 40 or 50 years ago the docks belonged to individuals but then government took charge and improved them. An argument in favor of government ownership is that if we decide to have clean good management of these ports and harbors, we should have government management. In this country, so far as fiscal principle is concerned, where the government runs this business our policy is to run it on the principle of fees. We issue bonds and we lease out the docks to private companies at a figure which will pay for all running expenses together with the interest. Although we have revenues amounting 5 to 6 millions of dollars a year, these are counterbalanced by interest charges. Other cities in this order are Seattle, Baltimore, Los Angeles and Philadelphia.

Exchanges and Elevators: Wheat elevators. In this country, North Carolina is an example. This is more important in Canada, which dates back to 1912. They have five elevators, and the revenue is quite considerable. However, in this country the prejudice of the American farmers against government business of any kind is expressed in our taking up this subject. A few examples we have in North Carolina indicates that instead of having business management we have political management.

## 22. Roads, Canals and Railways

Another stage of government in business is transportation of persons and freight. First roads, the high roads. In the case of bridges we are in the transition period, originally sources of revenue, under the later system, they have tried to break even. Then the pressure was brought for the principle of deficit, so the bridges have become matters of expenditure.

Canals are also in a state of transition. Private canals became public canals and were sources of immense revenue; when railroads began to compete with them we followed the principle of fees. The Panama canal is about the same, though the tolls hardly pay the current expenses. The greatest canals in the world are government canals but run under the principle of tolls. Most government canals run under the system of deficit. The whole canal era in this country is a thing in the past. Yet on the other side of the ocean we find a renaissance of canals. This is one of the most complicated questions, e.g. what shall be the principle of charging for canal service? Whole books have been written about this. It is not a practical question with us, because our few canals follow the principle of gratuitous service.

Ferries: Most of the ferries are municipal ferries. Most of the private ferries run down and become municipal ferries. Immense sums are spent upon them, and immense employment. Government service is very much more efficient than private. On the other hand, this was not decided upon the principle of deficit, but we find that we lose money on them.

Railroads: In economic and political science this is one of the most discussed questions. We have examples throughout the world, private railroads and government railroads. When it is a government business, as in Russia, they try to make money. Some other governments, such as New Zealand, try to break even, and other governments, such as Brazil and Spain, lose money. Our experience shows that we should not have railroads controlled by the government. The arguments in favor of government ownership are as follows: (1) Railroads are common highways, king's highways. (2) A government railroad would do away with the considerable evils of private railroads, which show discrimination, not always designed but sometimes undesignedly. (3) Low rates, as all profits will not go to railroads. (4) Labor conditions should be far better. {They felt they could do much better to deal with the government.} (5) Social arguments. Railroads after all are private railroads and are for profits, whereas government railroads are social utilities. (6) Government could utilize the railroad system. In Germany there are a great many arguments in favor of government ownership; most countries have adapted it. In Canada they desired government ownership to save money. On the other hand, the arguments against government ownership in a great many countries also have been equally strong, and in some cases stronger.

There are three classes of arguments: Economic, Political and Fiscal.

Economic Argument: That we are dealing here with immense capital outlay, a most difficult and huge business in the world, and, other things being equal, governments are not fitted for it. Furthermore, it is a most complicated business. Railroad presidents get more money than our presidents, because they require able men. Also, while it is true that we would not have any discrimination, yet there would probably be very great compensating economic disadvantages as regards profits. It is very doubtful whether in this country government ownership would mean lower rates, because of the inefficiency of management and greater cost involved.

Political Arguments: They are still stronger. Millions of people are directly or indirectly connected with the railroad business. It would bring about demoralization, and, so far as rates and fares are concerned, how is it possible to keep different interests from getting their fingers in the pie?

Fiscal Arguments: In a country like this with no budget system, to pitch further the expenses of railroads into our whole revenues would immensely complicate our budget. It is very difficult now to keep up with our system; adding railroad expenses amounting to billions of dollars would make the situation immensely more difficult. Our states and cities would be deprived of the revenues that they get from the railroads. Government ownership of railroads in this country seems unthinkable.

Transmission of Utilities and Power. There are two general classes: (1) State enterprise, (2) Municipal enterprise. The only examples of state enterprise on this Continent are in Canada, especially on their side near Niagara Falls and the Winnepeg Power Commission. In this respect Canada is more advanced than the United States, where government enters into the electrical business, engineering business, etc.

## 23. **Municipal Monopolies: Water, Light and Power**

If you take these businesses you find that in every case they present the following four characteristics: (a) Public highways. (b) Almost universal use. (c) House to house service. (d) The business is inevitably a monopoly. In the case of water we are dealing with a fairly simple business; in the cases of gas and electric light and power, everybody is interested in them, and the business is more complicated. Almost all of these businesses follow the principle of increased cost. Further, they deal with necessities of life. In the case of water works, statistics show the tendency to have been distinctly in favor of government management. But in the case of gas only a very small percentage, around 6%, is in the hands of government. Philadelphia is one example; also Richmond, Duluth, Wheeling, Ohio, etc.

England is the opposite, with much more progress having been made there. There are few private gas businesses in England. It may be because of proximity of the English cities to coal supplies. England is great example of municipal gas works.

In the case of electric light and power, the situation is different. Electricity started in the United States and almost all of our towns have their own electric plants. In 1912, there were 1500 municipal electric light works, and 3600 private works. Only 5% of the power was generated by government, and out of 62 cities you find only 8 with municipal plants.

Street-car lines are rarely operated by municipalities. San Francisco and Seattle were the first big cities to take it over. The subways are to a certain extent owned by the city. In European countries almost every city owns the medium of transportation.

Philosophy of this: What is the expediency of state action? Is it good or bad? You must remember the criteria and must distinguish between different countries and cities, the character of the enterprise, the fiscal situation, the social situation and the political situation. There are arguments for and against municipal ownership.

Arguments against municipal ownership: (1) Immense expenses are involved, which amount to hundred of millions of dollars. (2) Politics enters into operation and consequently bad management follows. (3) The labor question, what are you going to do in case laborers strike? (4) The constitutional question, how to pay in case it is desired to purchase a private enterprise. (5) On the fiscal side, the entire system of taxation has to be adjusted. Government will have to run the operation at cost. These arguments are very strong.

Arguments in favor: (1) It might reduce prices but the great point will be its social policy and not private or business policy. (2) Social principles are much easier to adapt with government than with private individual. (3) The treatment of employees will be somewhat better. (4) The chief argument is that the public will welcome the abolition of the political interests involved. You cannot get away from the political system in any way, whether private or public. In Philadelphia when the gas was in the hands of private people, it was poor and expensive. The movement in favor of municipal ownership is weaker today than it was 10 years ago. The subways of New York City, for instance, should operate without having a gain in view.

Fiscal Principle: If run as a municipal enterprise, according to what principle should it be run? In Europe it is run by the principle of benefit. In this country it is quite safe to say that public opinion will not permit the running of municipal enterprise on the benefit principle. Taking water rates, although theoretically we pay our way, yet as a matter of fact, taking into consideration the loans we have made, they have not only approached the fee principle but have come to gratuitous service. There was a plan to give water freely. This is not possible as it will very much increase the use and will be unconstitutional.

The principle that will be followed will be somewhere between the principle of fees and the principle of gratuitous service, which means vast increases in taxation. In the matter of water supply, some very interesting problems are seen, for example, the question of flat rate, cost rates, weight rates, service rates, etc. What is true for the water business is also true for gas rates, namely, different types of rates: commodity charge, customer charge and the demand charge. A whole science of proper rate making has arisen in the gas and electric light and power businesses.

## 24. **Public Prisons and Workshops**

This only incidentally touches fiscal science. The problem is different here. The first reforms started in this country about 100 years ago. We have gotten so far as employing the convicts in order to derive benefit. This may be divided into six classes: (1) The lease system, where the convicts are leased to private individuals, such as in Virginia, Wyoming, etc. (2) The contract system, where the state gets the revenue. (3) The piece price system, where the state itself maintains these institutions of work and supplies their raw material but does not always manage the whole business. (4) The public act system, where the state does manage the business and assumes the expenses. (5) The state use system. (6) The construction and repair of public buildings. Although we get a revenue from prisoners, the revenue is more than outweighed by the expenses.

## 25. II. **Fees: History, Classification, Description, Criticism**

The history of fees is interesting. They are found among the earliest forms of public revenues. Fees are paid for two purposes by the individual to the government for something done by the government to him. The government does two things: (1) Actually undergoes certain expense in performing something for the individual; (2) The government, after having declared a certain process illegal, may give permission to the individual to do what otherwise would be illegal. What is the difference between prices and fees? In the case of price, the government does something to the individual, and although the thing done by government is no longer considered to be an ordinary business, the government does not give the water or gas away but charges something. If it becomes a government function, we then speak of the government institution as protection, street cleaning, etc., and call it a fee. When we are dealing with a government-run business, the charge is a price. This is a distinction more of degree than kind. A fee is something that is paid by the individual for something of a relatively small outlay. Another class of fees is payment for privileges. The extent of fees of course varies considerably in amount during the past century. They were paid by individuals in return for banquest [sic] and in most cases developed into indirect taxes and high burden, but nowadays fees are limited to the amount of cost, although very insignificant, yet altogether they aggregate to a large amount. There is no way of determining the amount of revenues from fees. In some cases we find they are paid to the municipal or federal government, but in almost all cases the fees are kept by the official who does the service and there is no telling what the fees are. In the City of New York fees to the Sheriff amounted to hundreds of dollars. These have been turned over to a political party. In the federal government, nowadays consular salaries are large, but a few years ago they were partly paid by fees; now these fees are returned to the government. Consequently millions of dollars are paid in fees but only a small proportion figures in the budgetary accounts; here some field reform is necessary in the near future. Fees are paid by the individual generally each time the service is performed, but sometimes when the service is done frequently the fees are combined. There are three great classes of fees:

(1) Judicial, connected with court costs and charges. Adam Smith still believed in paying judges by fees. In certain places, fees are pretty high.
(2) A class of fees connected with business. These are fees paid for the privilege for carrying on a certain form of business, chiefly in our city, pushcarts, licensed hackman, and auctioneers. The problem is whether the fee is to be considered a fee or a license tax. The real difference is in the extent of payment: if it is small we call it a fee; if however, they become very high, then we call them license taxes.
(3) A very large class of fees is connected with an administrative service. Sometimes they are classified according to department of the federal government, for example, Secretary of Agriculture, Secretary of Commerce, etc. But it is much better to classify them according to the criteria of fees themselves. For instance, there would be fees of permits – 50 to 100 classes of permits – such as carrying pistols, maintaining a stand near a shop window, permits for building, etc. Wherever there is some desire to regulate public activity we issue these permits for special privileges as public concessions. We get great revenue from patents and copyrights; in the case of changing your name, the fee in this city is high. We might include registration fees, such as birth, death, marriage, divorce registration, etc. We also have commission fees: public commissions, notarial commissions who attest ail sorts of documents. Far more frequent are the fees paid to technical commissions, bank commissions, public service

commissions, all sorts of commissions instituted to regulate things that are all-important to the public. All these commissions are supported by their beneficiaries. We also have educational fees, which aggregate to a certain amount but are not very important; admission fees to museums of arts and botanical gardens; and all manner of miscellaneous fees for all kinds of things.

From the point of view of the individual, fees amount to a large sum, but from the point of view of revenue to the government and also from scientific point of view, their importance is much less. The only problem is, who is going to keep these fees? This is not a scientific problem. We now charge a very high fee for passports, ten dollars each, and each European country followed this example. For a traveler it is rather a big sum to pay ten dollars every time he crosses the boundaries of different countries.

## 26. III. **Special Assessments: History, Theory, Purposes, Methods, Extent, Criticism**

This is a peculiar development. It is a payment made by the individual to the government in return for the improvement to his real estate. A special assessment therefore differs from both tax and fee. A special assessment is not a tax, because it is paid for a special benefit that the land gets from this and taxes are paid not for any benefit. On the other hand, the special assessment differs from the fee. It is paid only for an improvement of real estate whereas fees are paid for anything. Special assessments are paid before or after service. Fees are paid for additions to the capital account for building new streets and sewers, while fees are paid for ordinary expenses. Special assessments are such an important class of fees that they deserve special attention, although they lie between prices on one hand and fees on the other hand.

Special assessments originated in Holland, where they were used for improvements. When the Dutch came to this country they used it in this city; throughout the 18th century we got most of our revenue from special assessments. This spread first to Massachusetts and is now followed in a great many states. Some use include: building advantages, pavement. putting out trees, building parks, etc.

The great problem in connection with special assessments is, can it be used for subways? In any rapidly growing city, revenues from special assessments are greater than taxation. This is especially true in the Western cities. In this city the revenue we get from this source runs to 15 to 20 millions of dollars. It is a convenient way to pay for expenditures. This was tried in the police department, upheld under the power of taxation. But they always distinguished special assessments from taxes proper. The difference is that in the special assessment there must always be an assessment area. The questions are, how far do the benefits go? and how far as a member of an assessment area do you pay this amount? In California a great deal has been done in the irrigation field. The theory is very simple. If this improvement is made and costs a great deal of money, how shall it be paid for? To put it on the tax bill would make it impossible of prosecution at all; to charge for it by fees or tolls used to be done in the Middle Ages, but is not done at the present time. If therefore you don't pursue the policy of fees and could not pursue the policy of taxes, what more sensible thing is there to approach the value of the property of those who benefit? This is a simple administrative arrangement.

# BOOK II

## Taxation – General

### 27. Definitions and Terminology

Some definitions are very long, whereas some are very short and inadequate. The best definition will contain the necessary ideas in the fewest words. The nearest is as follows: **A Tax is a Compulsory Contribution From the Person to the Government to Defray Expenses Incurred in the Common Interest Without Reference to Special Benefits Conferred**.

The historical development of taxation has been under the following heads: (1) Benevolence; (2) Precaria; (3) Assistance, aids or subsidy; (4) Sacrifice – to give up something; (5) Obligation; (6) Compulsion – as impost; (7) The idea of rate or assessments.

We speak of payer and bearer of taxes. The base, the thing on which taxes are assessed, the general unit of base value is dollar. The tax rate is the amount falling on a unit. This is the general rule. In some states, however, the unit of the base is not a dollar. In Vermont they make up only 1% of the amount of property. We speak of the tax list, tax book and other words.

### 28. Classification of Taxes

There are three different kinds of classification: I. Very obvious criteria. II. Criteria which are non-exclusive or of very little importance. III. Really important criteria.

I. Example of obvious criteria: (a) The property in which the payment is made, taxes in kind and in money. (b) The criteria of frequency of payment, or of ordinary and extraordinary (war) taxes. (c) The purpose for which the taxes imposed, is it a general or special tax? (d) The criterion would be jurisdiction: local, state or general.

II. Non-exclusive criteria, or criteria of little importance: (a) The source from which the taxes are paid (rents, benefits and wages in the ancient times). (b) The mode of measurement, such as property taxes and income taxes. (c) The period of intervention (assessed taxes and expenditure of taxes). (d) The character of the economic phenomenon: taxes on possessions, taxes on transactions. (e) The legal criterion: taxes on persons, on property and on business.

III. Really important criteria: (a) The criterion will have to do with the rate. Is the tax perpetual or gradual? Is the rate the same or is it graduated? And, if it is graduated, is it graduated up or down? Is the graduated tax progressive or regressive? If you look at it from the bottom it increases, but stops at a certain point. We also have degressive [regressive]. (b) Method of payment: apportioned and percentaged taxes. Apportioned taxes are arranged as follows: First, how much expense will be necessary? Say, two millions of dollars. This is divided into different counties. The county heads divide this sum among the towns, cities, villages, etc. They know about how much it will be and add to this the expenses. If it amounted to three millions of dollars, they take the assessed value of all property and divide one by the other and they get the value. The percentage system is seen in the income tax, such as one, two, or three percent. (c) The result of payment is often distinguished as direct and indirect, used in their modern sense. In the 18th century the general opinion was that all taxes would fall on land; that being so, the only direct tax should be on land. When the Physiocratic theory of distribution disappeared the distinction between direct and indirect tax was no longer with regard to the land but who bears the tax. Where the payer and bearer of taxation are different persons the taxes are indirect. (The trouble with this distinction is that when you import silk the commodity is taxed but if you sell it or use it yourself there is some difference.) John Stuart Mill says it all depends on what legislators intend to do. But the trouble is we do not know what the legislators intend. The French opinion is that it depends upon the nature of the payment. If you are dealing with a durable object, it is one thing. If you are going to make a list, it is a direct tax; if you are not going to make a list it is an indirect tax. This is also inadequate. The administration says: What I call a direct tax is direct, and what I call an indirect tax is an indirect tax. But there are taxes which are neither direct nor indirect, such as inheritance taxes. What is a direct tax? What is an income tax turned upon inheritance? We do not know. A direct tax in New York means only the land tax, the real estate tax; any other kind of tax is an indirect tax. To make this a scientific classification it is best to combine John Stuart Mill's definition with the French definition. (d) The criteria would be stages in the economic process of wealth. It has got to be acquired, possessed, assessed and then consumed. Some writers say acquisition, exchange and consumption. (e) The criterion would be when you take the mode of levy or degree of personalty which enters into the tax. This receives no attention in the literature but is most important in our day. You have to distinguish between a personal and an impersonal tax. The personal tax is a tax on the person. Of course, all the taxes are upon the persons, a poll tax, income tax, property tax, etc. However, over against that you have impersonal taxes laid on commodities, such as automobiles, cigars, transactions, sales, transportation, or communication. Those are all impersonal taxes. It does not make any difference who is the owner. In between these taxes which are purely personal and purely impersonal there are taxes upon things in reference to some person, for instance, a tax on land in the city. The tax is not paid by the land, a man has to pay it. Or, a tax upon personal property owned by an individual, a tax upon a mortgage. The man owns the mortgage is going to pay. Or a franchise of corporation, or business tax. It does not make any difference who the individual is, whether rich or poor, whether he has any children or not, the tax upon a business is paid by the business. This I would call a semi-personal tax.

I suggest semi-personal tax because the tax still is payable by a particular person, yet no reference is made to his individual income tax. If it is a semi-personal tax no attention is paid to his other income. In Europe a new tax was developed in the past two years which is quasi-personal. In England when they needed more money they went direct[ly] to a system of personal taxation. In France, when the will of the Ancient Regime was still very strong, they developed a new income tax. They have the tax first on the income and in addition they have the tax upon the incomes of particular things.

What is the difference between our way of land taxation and the French scutage tax? Here you have three different things. In the case of income tax you can make deduction for one's children, etc. In France the tax is assessed upon land irrespective of the personal condition. It is half way between: The tax is still upon land, but certain allowances are made for debt, etc., but they don't go the whole length, they have no graduated system. The French and Italians have something in between semi-personal and personal taxes. What are you going to call this? I would call it a quasi-personal tax, which seems to be a rather subtle idea. Classification terminology is a very important thing. We must bear in mind the distinction between impersonal, personal, semi-personal and quasi-personal taxes.

## 29. Historical Development of Taxation

This is such a wide field that one can spend a year lecturing on this subject. However, we can only call attention to a few main points. They are: (1) Political and (2) Economic. Under the political we have (a) War and (b) Liberty. Montesquieu said that the amount of taxes depend upon the amount of liberty. Free people are willing to pay a great deal more than slaves, and they are more able to do so. The economic influences are: (1) Stage of economic progress of primitive community industries and the simple or single tax system. The more you develop the economic situation the more complicated your economic life becomes; that is why our tax system is complicated. (2) The relative importance of the factors of production. (3) All economic life starts out as local life, then we gradually have state taxes, then federal taxes and we are going to have international taxes – League of Nations. (4) The forms of business organization. Business goes through many forms: individual business, the guild system and, in modern times, partnerships and corporations. (5) Development of the idea of ability to pay, which has had a very slow evolution. The personal tax comes nearer to ability to pay. (6) The importance of administrative methods. In early times they got returns from taxing the farms and gradually added and administered all taxes. (7) The change in the measure of wealth: in the past, in trade and stock and capital; nowadays, in terms of benefits and income. (8) Emphasis on social considerations. We sometimes talk about economic institutions, as political and social economics. All these points influence the development of taxation. If you bear in mind the things mentioned, you have the key to the whole situation.

## 30. Ancient Taxation

This is of very little consequence, as taxation played a very small role in Greece and other ancient countries. In Rome, however, you have the development from peasant to the greatest empire ever been seen. You have all sorts of development ending up with the greatest abuse when everything you can think of was taxed and when torture was employed in order to make a man disclose his property. This shows us what to avoid, and what to learn. With one exception the genius of Roman Law differs from the mediaeval system; in theory it was the principle of ability to pay.

## 31. **Mediaeval Taxation**

England is a great example. You have both impersonal taxation and indirect taxation, a land tax and then, after industry comes in, they develop what we would call today a general property tax. Only in England at that time land was never bought and sold; therefore you could not levy any tax on selling price. Consequently they got into the habit of taxing the rents of land; then they taxed personal property. Thus the whole mediaeval system was "10th and 15th". The system was a percentage system. After the system broke down, instead of taxing every man, they apportioned the tax to each county and, at the end, 10th or 15th meant so many pounds. Even this did not work and the whole system was abolished. It started out as a land tax but when personal property developed it became a general property tax, 10th and 15th. Instead of being a percentage tax it became an apportioning tax; then finally it went back to being a land tax and personal property escaped entirely. They tried this system time and time again. By the end of the 17th century it was a land tax. In the 18th century it became a rent charge, which is the amount of money charged upon the rent of land. You could pay a nominal sum to get rid of the entire tax. When England needed more money they tried the taxation of business; during the French Revolution they went to an income tax. At that time a man's prosperity was measured by his net income. What took place in a broad way in England took place all over the Continent. In France it was called *taille*, which was nothing but a general property tax, the strong man to bear the burden of the weak man. Under the influence of the military system France gradually introduced the system of exemption of noblemen, clergy, and lawyers. This was the greatest example of the inversion of ability to pay. When the French Revolution came along they wanted to get rid of all taxes, indirect or impersonal, and introduced what we call a quasi-personal tax. Only a few years ago they came back to a personal tax. In this country in our colonial period we had three systems of taxation. In New England, it was primarily based upon ability to pay. In the Southern States, it was just the other way around; they had lands, slave owners, and had very few poll taxes. In the Middle States – the trade center – they had a different system. Since business was the most important thing you find taxes upon business; excises and general property tax didn't develop until very much later. The tax system in Rhode Island was like that of New England, whereas New York was Dutch. So that you will find very excellent examples of the influence of these differences in our own history.

## 32. Modern Taxation

Modern taxation is marked by five characteristics: (1) Up to the 19th century there was a great swing toward semi-personal taxation. Since the end of the century the great swing is toward personal taxation. (2) The great modern development has been the substitution of direct for indirect taxes – the ability to pay. (3) A modern development is the swing toward administrative improvement. (4) The adjustment of the modern industrial condition: With the development of corporations, the corporation tax becoming both the most important and the most defective. (5) The influence of social reform ideas.

## 33. Local Taxation

This is the characteristic of the modern life. The immense development of local functions and therefore the immense increase of local expenditures for better schools, roads, etc. in civilized life. This is local taxation and as the proportions of these functions grow broader and wider they bring up the whole problem of local and state finance. How to raise millions of dollars and build bridges, roads and other improvements etc.

## 34. Essentials of Taxation

What are the essentials of taxation? Among them the important things are:

## 35. Reason of Taxation

The philosophical justifications of taxation: we have had two ideas, the contract or exchange idea over against the obligation idea. We base the doctrine of obligation of people to pay taxes simply upon the fellowship of the community. The real philosophical justification of taxation is found in the fact of fellowship.

## 36. Function of Taxation

What are you going to do with taxation? There are two contrasting theories: You might call these theories (a) Fiscal theory and (b) Sociopolitical theory. The first is represented by Calhoun or by D. A. Wells. Wagner says in taxation you go through several stages. You start in the narrow state and go on to sociopolitical purposes. The benefit lies in between. There are two points of view: (a) No matter what you think of it there are always going to be some social effects of taxation. A tax can never have a purely fiscal result and it is better to have some social result. (b) Apart from the fiscal result government has always utilized taxation for certain social purposes and always will use taxation to foster industry. We also use taxation to prohibit the consumption of opium, to prevent child labor, etc. Wherever public opinion has got to a point that the main social result is desirable, they use taxation to that end. You can say that the function of taxation is to raise revenue, but in such a manner as to harmonize with the social ideals of the community. Another point is, what is the purpose of taxation? Entirely apart from the social aspect, everyone agrees that taxes ought to be levied for the common benefit in this country. We have had controversy over the question of taxation for public purpose. All judges took a hand in that controversy.

## 37. **Nature of Taxation**

Is it a good thing or a bad thing? Again you have different points of view. It all depends upon the distinction between the production and consumption theories of public finance. The distinction between production and consumption is by no means anything but utilization. Most writers distinguish between production and consumption. Advocates of production say taxes are a good thing, consumption theorists say taxes are a bad thing. Some say taxes are spent for common consumption but it is not productive. Both extremes of the points are not sound. The mistake lies in the confusion between economy and niggardliness. Reducing taxes is just as bad as over-taxation. The point is that everything necessary is good. If taxes in that sense are desirable, then the question arises, what is the limit?

## 38. **Limits of Taxation**

You have to discuss taxes from points of view not in relation to rate but in relation to origin of source. The question of tax limits necessarily brings the problem of the question of social income. If taxation can stand the social strain it is all right, but if it cannot the dangers are great. And there has got to be a surplus over cost. If the tax is too much it either retards or entirely diminishes prosperity. Where taxes are very high, the danger to enterprise is enormous, just as in Russia, who wants to borrow money from this country. They should first show that there is a certain surplus over expenses in their country. The limit of taxation is that point where it bears a certain reasonable percentage of social income, where it does not succeed to check or destroy enterprise. Beyond this point it is suicide. There are constitutional as well as social limits. By the economic limit, you must keep taxation in a certain relation to the source.

## 39. **Source of Taxation**

What is the real source of taxation? Of course in a certain sense you can find that the source of taxation is the source of income. Income is not everything that comes in. Income is net income over against gross income. It is the surplus. Income is the surplus over expenses of securing the income. Net income is one thing and gross income is another thing, and if you are dealing with a taxpayer you must differentiate his net income from his expenses of living. If you tax and it takes away the whole of the surplus, the taxation is excessive. The man is going to live, but to take the whole net income is absurd. You must contrast the net income with clear income. In the case of the individual, the amount that the government takes in taxation is in a certain relation to net income, but in the case of society the only concept is the social clear income. We measure by taxable income what approaches the concept of clear income. The matter of sickness and disability has to be considered. This is a very difficult problem. The difficulty arises as follows: by income, do you mean money income or psychic income? Man renting an apartment, paying $2000 a year for it; and another man lives in his own house, does not pay any rent. Is this income or is not income? Income according to some people is that which comes in regularly from a definite source. Is application of capital, income? Or, is it capital increment? Suppose there is a change in the value of money, you buy a house the price changes, is it income? Or, are stock dividends income?

## 40. **Incidence of Taxation**

When a tax hits someone, it is called inflicting the tax. If the person who pays the tax is able to shove it off, the process is called shifting. When it reaches the person bearing the tax, lying upon that person, it is called incidence of taxation. Shifting is only one way you get rid of a tax. There are a great many other ways to get rid of it. Smuggling is one way, reducing consumption is another, etc. But if you continue with the subject matter, then of course, there are still other ways of avoiding tax. One is amortization; another is capitalization of the tax, as the price of a bond will fall on account of capitalization of tax. Transformation of taxation: Shifting is mostly used for getting rid of taxation. It is only by careful attention to these principles that taxes can be shifted. The Physiocratic doctrine was interesting but unsatisfactory because their theories of distribution were based upon the fallacious theory that everything was produced by land. Adam Smith or Ricardo followed their methods in distinguishing between the direct and indirect taxation. Ricardo had something to say about shifting taxation. What he says about this matter with his general theories of economics will bring out the distinction between social phenomena, what he calls the residues and the incidentals. It is the fundamental and the residual things. What Ricardo said is to take the fundamental thing into consideration but discard the incidental. Now in between these there came another theory which held the ground for a long time. A Frenchman, Gonnard,[166] said you must look at economic life. In the human body as well as in a tree there is circulation going on all the time; just as the blood will carry anything injected all over the body, so does taxation. Thiers took up this theory as a weapon against socialism, saying that it is foolish to suppose that any kind of tax will do. He said it resembles rays of light; a tax once applied will at once be carried all over the body economic. He called this diffusion. This theory is accepted even at the present time. This, however, applies to a little part of the problem and it is not sufficient to govern. The modern theory attempts to explain this more fully. Since you are dealing with taxes and with value you better consider consumption and production. Some consider both but some of them do not. In the first place, let us suppose these questions: (a) Whether you are dealing with monopoly or competitive system. If you have absolute monopoly, the tax cannot affect the price. (b) Distinction between general tax and exclusive tax. Property tax etc. Most taxes are special or exclusive taxes. (c) Is there mobility of labor and capital? The more mobile your phenomena the greater the chance for spreading the taxation. (d) Whether the tax is large or small. (e) Is it a new tax or an old tax? These apply for both.

When you are dealing with a tax on something which is demanded everything depends upon the elasticity of demand. Some commodities are more elastic than others. The more elastic your demand the easier to shift the tax but in the case of supply there is also a very interesting consideration. It all depends whether tax is imposed on the marginal supply or on the surplus. It also depends upon whether there is a sellers market or a buyers market. Sometimes we have sellers market, sometimes we have buyers market. Equally important is the question of elasticity of supply which is a very complex problem; for instance the elasticity of supply depends upon the principle of the ratio of product to cost. Decreasing returns or increasing cost is the ordinary thing. If within a smaller output it will cost less than before, how far it is shifted will depend upon the elasticity of demand. If you sum up all these considerations, you have as a result a series of rules. There are ten or eleven rules as to the conditions under which you can shift the tax. What is more interesting to us is the practical application.

Taking up the chief practical problems, the most significant is a tax on real estate as between (a) a tax on the farmer (b) a tax on the city man. Some say that taxes on the farmer are defused, but the more careful question is whether it is a general tax or an income tax. If it is a general tax then, of course, the tax cannot be shifted but that is a theoretical proposition. In the case of a special tax exclusively on real estate, the first problem is whether there is absolute mobility. Is the tax upon the margin or upon the surplus? Is the tax large or small? If the farmer produces things that are consumed in the same locality it is easy to shift it, but if he produces an international product then he has no control. And if he is a marginal producer, he will find very great difficulty in shifting the tax. So it depends upon all the problems of international supply and international demand. In a general way it may be said that the tax upon farming land being a special tax, being an exclusive tax, being an international tax in raising cotton and wheat, there is great difficulty in shifting.

When you come to city land, the situation is still more complicated. The city tax is a tax on real estate, such as buildings. There are immense changes in the improvement of the land. Population increases, the value of land goes up. It depends upon the borough, when you are dealing with different part of the same borough and different [things?]. A house is capital and an exclusive tax on capital is always shifted. Whether the tax on real estate will be shifted or not depends upon (a) Is it a general? (b) Is it mainly on land or mainly on the house? (c) What is the character of the locality? Is it progressive, stationary, or declining? The same thing is true in the City of New York. Some sections are growing in value and some are going down. Where population is increasing, a tax on houses can be shifted. If it is stationary it [can]not be shifted, and only a part of the tax is shifted to tenant. In some taxes some part and in other taxes a large part is shifted. [If] the tax on agricultural land is not general, it is not shifted; [a] tax on city land is shifted in a part, sometimes a great deal larger than the agricultural land. A much more difficult problem is the other taxes, taxes on different kinds of property, income tax, etc.

The shifting of local rates in England on real estate at least falls upon the occupier of the house rather than the owner. [A tax] on personal property and a universal tax on property is ideal but not a practical thing. We have to deal in this country with unequal taxes. There are three different points of view: (1) The relation between an owner and a new purchaser and a tax on bonds. (2) The different points of view of debtor and creditor. This is important in this country because of real estate mortgages. They are reached by tax assessors. If you have an unequal tax or an excessive tax on any one kind of property, then you cannot make the owner bear it. The tax will be paid not by the man who lends [but] by the man who borrows it. We had such a tax in New York and in California. It was shown that the tax was shifted to borrower, [not?] only the tax but they paid more. When in Wisconsin similar investigation was made, the rates did not go up so much. That brings up the general theory which depends upon several presumptions: (a) That you have a complete and free competition. There is in New York competition but not in remote districts, and the competition was not the same amount. (b) Free competition assumes knowledge by both parties; then you may have a degree of immobility. Then again the theory is that there should be no legal obstacles. So that while it is true in general that a tax on the lender will be shifted to the borrower, it is not always true under all conditions. We have to make a particular study for each case. In the City of New York people borrowed money to build houses where one could borrow 90% of the amount. This led to confusion. This is one phase of the question. More important is perhaps the application of the principle of the relation between producer and consumer, that is, where you are dealing with ordinary commodities which are sold in the market. A tax on property is a tax on benefits but unfortunately in many of our states we have taxes which are midway between taxes on the selling value and taxes on the net profits. Taxes sometimes are imposed upon gross profit or gross receipts; sometimes the taxes are lump sum, a flat tax, a fixed amount which does not vary. In each of these cases you have to consider the nature of the tax, which differs from case to case according to whether the tax is upon gross or upon net. Of course, the difference is that when you are dealing not so much with elasticity of demand but with elasticity of supply, the laws of increasing cost or laws of decreasing cost arise. In one case, it will be shifted and in another it will not be shifted. If you are dealing with a general and exclusive tax, whether you are dealing with all corporations or a certain class of them as some manufacturing corporations are exempted. And above all you have to deal with the question of monopoly. A few practical conclusions: Take a tax on the machinery of the manufacturer, under actual conditions such a tax ought to be shifted but the ultimate effect may be different. Take a tax on live stock of the farm, on the cattle, on the sheep or on the tools of the mechanic. The tax on the live stock on the farmer, other things being equal, will [be] shifted, after the changes have been brought about in farm tenancy. So that ultimately a tax of this kind must immediately be effected. Taxes on banks: If the tax is on the profits of the banks, the taxes might be shifted. Taxes on insurance companies will be shifted in the sense that it will lead to diminution [of] dividends to the policy-holder.

Now we come to a tax on corporations. Tn the case of public utility corporations prices are often fixed by public commissions. If the rates are fixed they could not be shifted but where rates are not fixed, if you have a tax on particular classes the tax will be shifted. In the matter of corporations there are two problems involved: A tax on the net profits of corporations is practically an income tax. A tax on income cannot be shifted but the more interesting question is the effects upon stockholders and bond holders. Stockholders are affected by dividends. The bondholders will not be affected, except that their security will be somewhat diminished. In the long run, selling value is the capital value made up out of present and prospective income. In the politics in this country a few years ago, and as to that matter even today, there was wide agreement that income taxes were bad things to the community [as] a whole, because they are shifted to the community. Income taxes cannot be shifted in England and in some respects that is true in this country. We have to bear in mind whether it is a biased market or not. In Europe the question of a tax on wages has been discussed a great deal. We haven't got this question. The ordinary [theory] in the earlier times, when people were under the spell of a theory of distribution of taxes, was that wages always would be shifted to the employer. It would be shifted if it was put upon marginal subsistence. The standard of life of the laborer however is now rising and the minimum is higher. There is always a margin between these. Owen D. Young[167] said that the minimum wage will be a cultural minimum. Certain professional wages have a quasi-monopoly, where you have a certain rule with reference to wages which will take the matter out of free competition so that it does not follow that a tax on a lawyer or doctor will mean higher rates to patients or clients. When you deal with such a thing as a poll tax and inheritance tax you cannot have any shifting, because there will be no one to whom to shift. When you come to a sales tax that has been a very active question. Senator Smoot[168] caused the sales tax to be killed by arguing that it will not only be shifted but more than shifted as there will be an additional charge, a pyramiding.

When you come to so-called indirect taxes the general feeling is more or less to argue about import duties. The tariff is a tax, but it does not follow that the tariff is always a tax. Theoretically and practically in particular cases, a tariff is not always a tax. You got to assume certain things about the market. The country may control the commodity. The importing country may be the sole market. We were for certain things the sole market in this country. Then the question of the elasticity of demand and of supply comes in. We might discuss other taxes, for instance, taxes on the stock exchange, about which there was a great fight as to whether the tax on the exchange is borne by the traders, members of exchange or speculators. The tax had various consequences. It was not borne in the main by the exchange as a whole but certain classes of brokers were driven out of business[;] in the [other] cases however the evil results followed. If you buy stock today you pay the tax yourself. The result was also seen upon banks and promoters. But that really leads us to the question of the effects of taxation. So far as shifting is concerned, it is a very complicated thing. Diffusion theory is not applied. You have to consider each tax by itself with these points mentioned. Some of them are shifted, and some of them are not shifted, and some less shifted.

## 44. **Influence and Effects of Taxation**

Even assuming that a tax rests where it is put, or even assuming that the tax is shifted, is that the end of the story? Not by any means. If the price of automobiles goes down 50 or 100 dollars, what effect does it have on the whole body economic? There are several different points to be emphasized: That of inequality or mobility of the tax. It used to be said by certain writers that old taxes are good taxes and new taxes are bad. In a certain restricted sense that is true. In 1857 a manufacturer of woolens petitioned not to have any change. People had accustomed themselves in this country to low taxation. Of course you can push this to an absurd extreme. There is a much more important way in which the inequality of a tax affects economic life. Consider the capitalization or amortization of the tax. If high taxes are put upon land in the city, if they are unequal, what is going to follow? You will pay very much less for the property as this brings down the benefits of the investment, so the investor has to charge more and invest such that he will get a certain return. An unequal tax upon any kind of durable property would be capitalized. If it cannot be shifted a tax on houses cannot be capitalized but a tax on the land can be capitalized. A new purchaser buys tax free and the original owner suffers the whole loss. This is called the doctrine of capitalization. Inasmuch as capitalization shows its influence, it shows through the change in the rate of interest. That is true of all taxes whether they are special or general. It will affect capital and will affect the rate of interest and will bring about changes in capital values. One can accept that criticism. That is to say, one ought to distinguish between variations in capital values and this particular phenomenon, which is due to inequality. The fact that a tax on land is capitalized, when other personal properties are not taxed, brings up the point that you are dealing with an unequal tax, as with a particular bond issue or a stock exchange when you are dealing with unequal tax. This doctrine of capitalization and administration deals with this inequality.

A second series of problems is connected with the question whether the burden on the individual is proportionate or disproportionate. Does the government get the money desired? The discrepancy, according to the burden and revenue doctrine, is that the government gets the amount less the cost of collection. Sometimes that cost is very high and is very important, such as in the case of the collection of the whiskey tax, whose cost of collection is more than the tax itself. There is another way. In the case of taxes on food and necessities, where a high tax would lead to shrinking of consumption, the loss to the individual is measured not by the tax which is paid but the tax which is not paid. There is a loss to the consumer and also a loss to the producer. There is another more interesting problem which may be called the excess-of-price-above-tax doctrine. Adam Smith points out that each time a commodity is sold the seller not only adds the tax, but also some more, and if this repeats the price of the commodity gets very high. You must distinguish between interest and profits; so far as the profit is concerned it is the marginal producer's profit. Insofar, however, as interest is part of the cost this is true, but the tax on the interest is small and even can be negligible. Ordinarily, therefore, when a tax is imposed upon a commodity there is no accumulation and there is no such excess on prices. But if you levy on an automobile, then every time [it] is sold the situation is different and there is a pyramiding of the tax and the increase of price. That is the reason why a tax on sales is not encouraged so much.

The third class of effects concerns whether the effects are intended or unintended. A tax on complimentary goods must be considered. Also, the question of evasion comes in, whether the evasion is intended or unintended, and the evasion of a tax on gifts and so forth is considered legitimate evasion. This is much more important than illegitimate evasion. This evasion comes about in the case of consumption, as for instance you can avoid a tax on tobacco or whiskey if you drink less whiskey and smoke less tobacco.

The fourth question deals with the effects of a tax. Are there any good effects of a tax? Does it stimulate production? Taxes sometimes improve production. This question we discussed under the name of transformation. The argument is, if natural difficulty makes you work harder, why would not an artificial difficulty produce more production? Wherever you find this, the tax is not the cause but the occasion to improve. Furthermore, over against this explanation, there are other ways that taxes sometimes do more harm. Henry George said all taxes except land tax are harmful taxes. There are four or five ways in which taxes do harm. The first the effects on production. High taxes of any kind may be repressive. If they are very high, they may render output unavailable; or they may be so high to leave very little inducement for fresh expenditure. Extravagant taxes may lead to wasteful expenditure; for instance, during the war, advertisement in the Saturday Evening Post was greatly increased, and the price of everything was low on account of excessive tax. High taxes will cause uneconomic changes in business practice. Taxes on sales in Spain have proven one of the worst taxes known. In Paris there is no real estate exchange because the tax on transfer is very high.

Effects on Distribution: Here we have a doctrine, "Leave them as you find them", called the ideal doctrine. But, this is impossible. It is possible if you want it, but most of the time you don't want it. Finally, you come to consumption and here there was a fight and the question was, Shall we have free breakfast table? Taxes upon savings are also an important question. Painful saving versus automatic saving is important problem in taxation.

The fifth problem is the indirect effects, rather than the foregoing effects of taxation. There is almost no end to the considerations that might be urged. You have to carry your analysis far enough if you want to get a satisfactory solution. Take, for instance, our real estate tax upon houses. A workman lives in a house, and taxes are raised upon real estate. What effect has this on rent and what effects has this on the wages of the workmen? What effect will a tax on stock exchange securities have upon a promoter? What effect will it have upon a security holder, or upon a railroad utilizing this means of raising money? In other words, when you take a big, broad view of the subject (which has never yet been adequately treated), you would have to say that there are two fundamental effects that taxation will have: (1) In some cases you have capitalization of tax, and (2) you will have division of tax. The conclusion will be that in the majority of cases you have elision, worked up into economic structure. In general it may be said that the difficulty of the whole problem arises from the fact that a tax represents a change in the economic equilibrium of prices. A new tax disturbs the equilibrium before it is taken care of and this leads to very complicated problems.

## 45. Canons and Principles of Taxation

By canons of taxation is meant the rules that should govern in working out taxation. In the older literature, rules meant the ethical rules about taxation. In the French literature on the subject, they meant by rules the administrative methods; they deal with the administrative side. Adam Smith combined both the ethical and the administrative concepts in laying down the principles of taxation. The principles of Adam Smith are as follows:[169]

(1) The subjects of every state ought to contribute toward the expenses of the government as nearly as possible in proportion to their ability, i.e. in proportion to the revenue which they enjoy under the protection of the State.
(2) The tax which each individual is bound to pay ought to be certain, not arbitrary. The form of payment, the manner of payment, the quantity to be paid, ought all to be clear and plain to the contributor and to every other person.
(3) Every tax ought to be levied at the time and in the manner in which it is most likely to be convenient for the contributor.

(4) Every tax ought to be so contrived as to take as little as possible out of the pockets of the people over and above what it brings into the public treasury of the State.

The following are very important criteria, and in a good system we cannot omit any of them:

First Criterion: Fiscal Criterion. The principle applied should be Productivity. A tax must be productive. Canons or rules: (a) Adequacy. Enough revenue to pay expenses. (b) Elasticity, because need varies from year to year.
Second: Administrative Criterion. Principle of Efficiency. If not efficient, it will not work. Canons or rules: (a) Certainty, you should know what you are expected to pay. (b) Economy, avoid extravagance. (c) Convenience.
Third Criterion: Social Criterion. Principle of Prosperity, i.e. to bring about the welfare of the people. Canons or rules: (a) Innocuous. Should interfere as little as possible with prosperity. (b) Harmlessness.
Fourth Criterion: Economic Alteration. Principle of Equality. Canons or rules: (a) Universality, a tax should reach all. (b) Uniformity.

The first three criteria are important from the point of view of the government. The last one is important from the point of view of the taxpayer. So, we call this the Primary and the others the Secondary Criteria. We take up first the Secondary Canons.

## 46. Secondary Canons

Fiscal Criteria. (a) Adequacy. A tax system must be adequate for the reason for which it is levied. A tax may not be called adequate if it violates some of the other principles, such as equality or some other. It must supply a satisfactory revenue without running counter to any of the other canons of taxation. (b) Elasticity: It may be defined as the responsiveness to the rate of yield [awkward: responsiveness of yield to tax rate]. If you double the whiskey or tobacco tax, for example, the people will refrain from smoking or drinking to a certain extent. So elasticity depends on the kind of a tax and the response of society to the tax.

Administrative Criteria: (a) Certainty. The best example is the difference between specific and ad valorem duties. Ad valorem is the tax assessed according to value, but who knows what the value is going to be in the future. You never can tell exactly what will be the value of a certain article. So in this case, the government levies a tax and is not sure what the tax will amount to. A specific tax, on the other hand, is definite, known to be a certain sum on the listed article. (b) Economy. This depends not upon the nature of the tax, but upon the social conditions of the community. When we had whiskey taxes, it cost $1\frac{1}{2}\%$ to collect in New York, but 35% to collect at moonshine districts in Kentucky. Sometimes the collection cost more than the amount of the tax itself. Concentrated taxes are more economical than diffused taxes. It should be the practice to levy taxes in such a way that very little need come out of the pockets of the people to pay the costs of collection and that the maximum yield comes into the treasury.

| Criteria | Principles | Canons or Rules |
|---|---|---|
| 1. Fiscal Criteria | Principle applied should be Productivity. The tax must be productive. | 1. Adequacy. Enough revenue to pay expenses.<br>2. Elasticity. Because need varies from year to year. |
| 2. Adminis-trative Criteria | Principle of Efficiency. If not efficient it will not work. | 1. Certainty – you should know what you are expected to pay.<br>2. Economy – avoid extravagance.<br>3. Convenience. |
| 3. Social Criteria | Principle of Prosperity i.e. to bring about the welfare of the people. | 1. Innocuity. Should interfere as little as possible with prosperity. Harmlessness. |
| 4. Economic Alterations as Criteria | Principle of Equality | 1. Universality – tax should reach all.<br>2. Uniformity. |

In the last lecture we called attention to the distinction between principles and canons of taxation and then passed primarily to those we call secondary canons. We said something about adequacy of taxation and also something about the elasticity of taxation as well as the absurdity. Now we shall discuss the problem of convenience to the taxpayer. The questions are: How are you going to collect it? When are you going to collect it? Where and under what conditions? For instance, how are you going to collect the tax? The question will be whether in money or in kind, or whether you are going to collect it by payment of stamps, as tobacco, or whether you will allow credit. All those questions depend upon the immediate environment, the economic situation, as to what is most convenient for the taxpayer. In the second place, when ought a tax be taken? Most taxes are bad in a lump sum. It is sometimes more convenient to pay in installments, as income taxes are paid in usually four installments, and in other cases more installments. The matter of convenience, however, is to be considered not alone for the needs of government but also for the convenience of the individual. Sometimes the government does not need all the money immediately; the matter of installment paying is as old as modern economic life itself.

Where the taxes ought to be paid, again, is not of very great theoretical importance. In the City of New York every tax is paid by check sent to the City Hall or paid in cash in different boroughs. In the case of Federal taxes we have collection districts; sometimes we have these in the states. There, too, the great point is to meet the convenience of the taxpayer. Perhaps the most difficult of these problems is the conditions under which the taxes would be paid. The land tax you pay in a lump sum or in installments, that is the end of it; but if you take the other extreme, the Federal income tax, there are a great many complications. The returns would be audited, visits made to your house or place of business, etc. In France it will be impossible to carry on the way that we carry on in this country, inasmuch as the French people would not stand for this; such was the case before the French Revolution.

All these points bring up the very important question of the convenience of the taxpayer. The next point is perhaps of little but not fundamental importance. This is the outgrowth of social criteria, such as we called harmlessness, that taxation should interfere as little as possible with economic prosperity. In our discussion of the effects of taxation we called attention to the limit beyond which taxation certainly does harm rather than good and to the repercussions on the body economic. All these canons are of a secondary nature and do not implicate complicated economic problems. But situations are different for the primary canons of taxation.

## 47. **Primary Canons**

These canons are the outgrowth of economic principles, the endeavor to secure economic equality among the taxpayers. The two canons which flow from this principle of equality are (1) the Principle of Universality, that everybody ought to be taxed, and (2) the Principle of Uniformity. Those are the important canons and around them center pretty much all of the significant discussions in the fiscal theory of modern times and all legislative fights. How are you going to reach both Uniformity and Universality? We shall now take up the problem of Uniformity and Equality.

## 48. I. **Uniformity and Equality of Taxation**

These are primary canons of taxation, but what do you mean by Uniformity and Equality? You can reach an approximate solution in a negative way, by deciding what you do not mean. You do not mean absolutely the same equality, as in the case of Mr. Rockefeller and a street sweeper. By Uniformity, therefore, you must mean relative equality. Then the question arises, what is relative? What is your criterion of Uniformity? You might say that all people with red hair, or bachelors, etc., would pay the tax; that is, of course, absurd. These would not be considered a satisfactory criterion. What is going to be the criterion of Equality? How are we to frame a system of taxation which will be Uniform and Equal? That brings us to the problem which has vexed thinkers for a very long time. What should be, in the first place, the basis of taxation? What should be the principles according which you are going to fix your cost? If we refer to the history in regard to this, we find three answers.

## 49. **Basis of Taxation**

For early governments, for that matter even today, the answers are as follows: The first answer is that taxes should be apportioned according to the cost of the service. If government does something for you, then you should pay something. The difficulty with this criterion is that, while it is entirely satisfactory where you can measure both the cost and the interest of every individual in that cost, there are a great many cases where you cannot do this. Where you can do this, the payment is not really a tax, such as when government runs a railroad and you buy ticket, or water rents, etc. All these things represent prices, charges, fees, but they are not taxes. The trouble is that when you come to the expenditures which are paid and taxes put, there is no way of apportioning the relative increase to every individual, such as when government builds a battleship, there is no way of telling to what extent this is going to affect the general amount of taxation. In the second place, it is a system which considers the value of service not the cost. In other words, the benefit that the individual gets out of this service may be more or less than the cost, just as taking dinner in the club house. Benefit is a closer approximation to the real basis of taxation, but here again we run against difficulty, because in a great many cases you cannot measure the benefit. What is the benefit to a bachelor from a school tax compared to the benefit to a man who sends 12 children to school? So that, in modern times, we substitute both cost and benefit with a new idea. The third basis is the ability or faculty to pay. This idea, of course, is almost as old as speculation. The Greeks discussed this, the Middle Ages had it, and during the

French Revolution they discussed this and laid down a regime for the payment of taxes according to ability. Even in the American Colonial period they applied the idea of paying taxes according to faculty. Then, we can ask, what is ability? or, what is faculty? Therefore, first, what is it?, and second, how are you going to measure it? Those questions bring very difficult problems.

What do you mean by faculty, or ability? This was not discussed until John Stuart Mill, who gave an explanation which had been widely accepted until very recently. Mill said that ability to pay is to be measured in terms of sacrifice. Equality means that the burden undergone by you should be equal to the sacrifice undergone by me. Faculty of taxation means equality of sacrifice. That was accepted all over the world until ten or twelve years ago. Two rather important thinkers took exception to Mill's doctrine: Professor Edgeworth[170] of England and Professor Carver[171] [Carr in original] of Harvard. They said the sacrifice about which we are speaking is not equal sacrifice but minimum sacrifice. {It must be such that in the case of any number of people there is always a smallest sacrifice beyond each and therefore they decide to substitute the minimum sacrifice instead of equal sacrifice.} The only way to impose the least possible sacrifice upon a rich man and a poor man is to take away from the rich man all his wealth. If you want to make the sacrifice minimum sacrifice you have to take a great deal more from the rich than from the poor man. In the case of Rockefeller's ten to fifteen millions of dollars yearly income, you have to make things very different. This means taking away from the rich man all of his surplus. This is called only a minor variation of the sacrifice idea. Mill's theory therefore is not affected by this. But there is another thing; the doctrine of sacrifice, whether equal or minimum, always has reference to the disposition of wealth, to the consumption. If you take out money from your pocket you cannot spend it for other things.

The real point of Mill is that it cannot be limited to consumption; there also has to be sacrifice in production, and sacrifice in acquisition. Adam Smith's statement, that the first 1,000 dollars is most difficult, fits in here; a small business finds it much harder to accumulate. You must work much harder in the beginning than in the end. That is what we mean by saying sacrifice of acquisition. It is much harder to gain a certain amount of money if you are at the beginning then when you are far advanced. So that sacrifice must be interpreted not alone in terms of consumption but also in terms of production. Therefore, we will say sacrifice liberally interpreted.

Another point is this: Henry George was a great man, but he was a one-sided man; though for all that George was a great thinker and introduced a new idea into philosophy and economic thought. The kernel of truth is the emphasis he put upon privilege. There has been and always will be certain privileges. The point is that this concept of privilege is one of the elements in ability to pay. There are all kinds of privileges, under the tariff laws, etc, so that, to the extent that privilege emphasizes the ability of individual, privilege must be considered an important factor in the concept of like sacrifice.

Now we are ready to consider what ability means. In considering the element of ability you have to consider sacrifice of: (1) Consumption, (2) Acquisition and (3) Privilege. This will tell us what you mean by faculty and ability. Now, how are you going to measure it? This brings us to the problem of measure, how are you going to measure one man's ability with another man's ability? Again, this is a very important and difficult problem. It is not that legislators follow the theorist, but theorists follow the legislators. If you consider history, you find about five answers to this question. Even at the present time you find the survivals of all of these five answers.

## 50. **Norm of Taxation**

What is the test of the measure of ability to pay? How does it work out? The first answer was, polls, take each man as you find him – "You see a head, hit him". Everybody pays the same, that was a great step forward in civilization, and wherever you have democracy you have a poll tax. As soon as you have more than heads, as long as you have heads acquiring wealth or property, then this becomes absurd, as it depends upon the economic characteristics of the individual. That is the reason why the poll tax has been abolished. Everywhere else in the world this is abolished; we find only traces of it in this country. Therefore the poll tax is not an answer now.

The second answer was expenditure. What a man spends was considered the best test of taxation. That also had a very interesting history in the Middle Ages. When you had an excess of taxation, the reformers said make all people pay. In England it was the great philosopher Hobbes who advocated that, and we find it frequently since then. The idea of expenditure originally was a perfectly legitimate concept but it was designed rather for the problem of exemption, in order to bring about universality. In the course of time the weaknesses of this idea became more apparent. If based upon consumption, the result would be to check consumption. That may be a very reasonable objective [objection in original], as in the case of war; they did so in Europe during the great war. That is very important in case of war but it is not important in normal conditions. Anything which checks consumption checks the fundamental cause of economic prosperity. The second objection, however, is far more important, namely, that a tax on expenditure involves a greater relative burden upon the poor than upon the rich. Although the poor man spends less than the rich man, he spends more in proportion to his income; at the bottom of the scale, people spend everything. So that a tax on expenditure is really a tax upon the poor. Above all, a tax on expenditure is out of line, according to our modern economic life. For that reason, in the democratic communities less emphasis is put on a tax on consumption; in the case of Cuba, for example, we did away with [the] consumers' tax. While this is not a good test of ability in general, we should admit that it may be arranged on special things, for instance, expenditure sometimes is a rough indication of ability, such as foreigners living in France. The amount of money they spend may in certain cases be utilized for tax purposes. Secondly, we may endeavor to primarily tax the luxuries of the rich. This was a favorite method in the 16th and 17th centuries. Thirdly, it may be desirable to restrict the consumption of certain things, such as opium. The object, however, here is not fiscal but social. Finally, a tax on expenditure sometimes is susceptible of being used to get rid of the difficult problem of double taxation. So that, in general, while expenditure is no longer legitimate as a general norm of taxation, it is defensible in particular cases. They tried polls, they tried expenditure, and in the civilized countries now they use property.

The third criterion, therefore, is property. This, of course, originally had a great many advantages. Property is measurable in terms of money value; as a consequence, we find this tax in all nations. The idea[l – ?] of taxation seems to be that general property tax, the advantages of which are obvious. In the course of time, however, evils appeared, so much so that today even in this country the property tax is fast disappearing. Why is this? There are five [sic] reasons:

(1) Property is no longer, as it was a century ago, a proper norm of taxation. The first complexity of economic life is the growing disparity between property and yield. Of course, ideally, this disparity does not exist. We know that [the value of] property, or capital, is simply the capitalization of income, so that there ought to be no disparity, but it does not always work like that. Take for instance, the farmer. The value of his property is the capitalization of what the farmer can get out of it. But suppose the flood comes along and profits disappear. The farmer has to pay taxes all the same. In the last few years this situation existed in some of our states; income disappeared and decreased. But in the meantime farmers have been compelled to pay [taxes] on their property even though their income disappeared. The same thing is true in stock and bonds. The Atcheson stock was a favorite investment and when Atcheson failed, the value of the stock was diminished but the owners of the stock had to pay on the value of

stock. With the more complicated division of classes in modern times, difficulties have developed.

(2) Brought to attention for the first time by President Walker,[172] taxing property is to impose a penalty on savings. Two men are in the same condition, and have the same earnings, one spends all he makes and the other man is economical and by self-sacrifice saves. The saver is taxed. Take two farmers, one spends his money on anything and the other spends his savings on painting his house, and consequently his house is reassessed and he has to pay more taxes. This makes for an inducement to spend rather than to save.

(3) When the property tax was developed, everybody was poor in this country. Nowadays a property tax exempts not only the ordinary laboring man but also railroad presidents and other presidents whose salaries range from ten thousand to sometimes 250 thousand dollars. The property tax was not calculated for these cases.

(4) The immense change in business. Some time ago the ordinary business man measured his business in stock and trade. The ordinary business man today measures his trade with what he makes and therefore a property tax does not represent at all the idea of economy. Two men with the same stock and trade may be entirely in a different condition at the end of the year. There has been a complete revolution in business transactions.

(5) A tax on property is like a tax on consumption. A tax on capital tends to deplete capital. That ordinarily is not a bad thing but in particular cases, as, for instance, with our forests, it is a weakness of our general property tax. A forest grower puts up more trees. After a while, from year to year, real estate becomes more and more valuable. A tax on property will increase the burden upon the real estate owner from year to year and finally lays an amount of tax that is unendurable and leads to cutting down the trees. This has been one of the greatest reasons for deforestation in this country. How to adjust these conditions is an important question.

(6) This reason is the most important of all the practical reasons. In proportion as economic life becomes more complicated, property becomes more and more intangible rather than tangible, as in the case of securities – stocks and bonds. Nowadays, of course, there are more intangible personal properties than tangible. It becomes almost impossible to reach the intangible. You can guess what a man's farm is worth, or his furniture, but certainly you cannot tell how many stocks and bonds he has, and that is the reason why the tax on personal property has completely broken down. The immense result of all these objections is that the extolling of property as a norm of taxation has disappeared and everywhere in the world property as

> the general test of taxation gives way to something else. In European countries this happened long ago and it is happening in this country. Property, therefore, is no longer a certainly adequate theoretical norm of taxation. But, just as we saw in the case of expenditure, whereas it is no longer dependable, it can be used in certain cases. These property norms can be used in minor cases.

What are these minor tests? (a) When the property is held for enjoyment. Take for instance, parks, where the property may be the only way of getting at the wealth of the owner. (b) When the earning capacity of a piece of property is very rapidly reflected in its capital value, then selling value is affected. In the cities we see that using property as a test for taxation may sometimes be legitimate. (c) When the yield of a piece of property stops, but its market value is still relatively great, then it may be desirable to impose a tax upon the property. It may be that its speculative value is very great, for example, some stocks may not have a yield but may have speculative value. In such cases this may be desirable. (d) A good many countries have utilized property as an extraordinary method of bringing about differentiation of taxation. In this country, as you know, we now have to differentiate earned from unearned income. In Germany they don't follow this, they tax everything; then they put on an additional tax. Property may be utilized as a method of bringing about differentiation of taxation. (e) War finance rather than peace finance. Imposing a capital levy is nothing but a single property tax. Sometimes this is 50%, 60%, or 80%. If you want a capital levy that is the only way you can get it, so that capital or property is still legitimate but can be used in special cases, and in general it is no longer found.

When property was eliminated as a norm of taxation, the next stage – the fourth stage – was reached when you took the yield or produce. The best example is during the French Revolution. Great abuses were connected with the taxation of property. In France they abandoned it completely and taxed only the yield of property. This was an immense improvement and spread from France all over the continent. The great advantage of the system was that it now paid closer attention to the real economic interpretation of wealth, wealth measured in terms of the yields rather than in terms of capital, or property. You could concentrate more closely upon the wealth, for instance, it paid no attention at all to the personal characteristics of the owner; and, just because of that fact, although it lasted a very long time, the growing democratic movement of modern times brought to the front some of the difficulties of this system. Two men own a farm each, 100,000 dollars each. Under the system of yield tax, both are taxed the same, [even though] one got the farm free and another man has to pay a high mortgage. Under the yield tax no attention is paid to this; when credit becomes more widespread, difficulties appear, one being the question of interest that must be paid out. Other matters also come in. For instance, one farmer is childless and the other is patriotic and has twelve children. This made the system so defective from the point of view of individual ability that it gave way to the final test that is widespread today.

The income tax is most serviceable; the advantages of the system are such that the income tax, or the idea of apportioning, has spread all over civilized Europe. In this country during the Civil War there was no income tax. During the last war we depended mostly on the income tax. This was a transition from property to income. While income is on the whole the best norm, we also have found it not free from difficulties. There are a great many shortcomings, not as many but still quite decided shortcomings, and these explain not a little dissatisfaction. What are these shortcomings? No two countries, no two scholars, answer the question the same way. There is no good book on the subject. By income is meant that which comes in. Among the problems of importance are: What do you mean by income? Money income, or psychic income? What is it? Money satisfaction or psychic satisfaction? Two men rent a house, one owns the house, the other pays 2000 dollars rent a year. Two farmers, one has to buy his feed, the other raises his own. Should any allowance be made for these differences? You will say yes, but it is not so [easy] in the actual world. If you say income means psychic income, you cannot carry the thing to the extreme. It is a very hard question to decide. Do you mean by income that which comes in money, or more easily calculable as one's money's worth?

In the second place, if by income you mean net income, net receipts, you mean therefore what comes in over and above expenses, but what are the expenses? Under this comes more problems, such as depreciation, amortization and wasting assets. This can be made solvable by accounting, so that this is not as difficult as the first. In the third place, is income that which comes in periodically, or not?

The third point under this head is the question whether you understand by income something which is periodical or otherwise; whether you take it from year to year or the average.

The fourth point under this heading is the regularity or permanence of the source, whether you distinguish between chance over against regularly recovered gains. If you make money out of transactions and securities, is that income? If it is your business, if you are a broker, is that income? If you go to Wall Street and make money, is that income? Is inheritance income?

All these questions are treated variously in different countries. So far as scientific investigation is concerned, it is to get away from the concept of regularity. Gradually we have taken in the irregular incomes. The desired tendency has been to broaden and also to include irregular gains, as well as regular gains. But the most important thing as to the meaning of income is to decide as to the distinction between accretion to capital and income. For instance, you buy a house, pay 10,000 dollars and sell it for 12,000 dollars. Is that 2,000 dollars income? You do not sell it, but the price rises. Is that an income? There, perhaps, has come the most interesting and difficult distinction and we have two extremes of views. You can take your choice.

First: Irving Fisher,[173] in his book on capital and income, has made a very interesting analysis. He makes a very sharp distinction between accretion to capital and income. Income is only that which when capitalized will be capital; therefore, if you count accretion to capital as income you are counting it twice. On the other hand, while a great deal of what he says is true, he goes so far as to maintain, agreeing with John Stuart Mill, that savings are not income. If a man has 10,000 dollars income, and if he saves 2,000 dollars of that, you must not count that income. Now, on the other hand, you have another point of view, first made familiar by a German writer.

Second: Chand[174] [Chanz in original]. His concept was elaborated by Professor Haig[175] [Hague in original] of Columbia University. He calls it the accountant's concept. He defines income as the money value of the net accretion to our economic power between two periods of time. That is an ordinary thing with business. You take stock at the beginning of the year and again at the end of the year, the difference is the accretion to the economic power of the corporation and would be called income by Professor Haig [Hague]. Professor Fisher decides that this is additional capital. So, in the same way as with the house, which was worth 10,000 dollars in the beginning of the year and 12,000 dollars at the end of the year, the 2,000 dollars difference is considered income, whether stocks are sold or not, the economic power of the individual is increased, as he can take it to the bank and borrow money. No, says Fisher, nothing of the kind; it is an accretion to capital. As regards the truth, it would seem to lie between the two.

Why, is Fisher not right? Because, if you exclude all savings from the concept of income, you get into an absolutely untenable position: "By their fruits shall you know them". It is theoretically unsound to transfer the idea of income from the source of production and gain to its consumption or disposition. What difference does it make what you do with it, it is income whatever you do with it. Practically, just think of the results: Here is a poor man, who spends everything, while higher in the scale the easier it is to save, in fact the rich man cannot help saving. If income excludes all savings and if the rich man saves 80%, or 90%, you see how absolutely absurd it is for the income tax, it is practicably untenable. If a theory is correct you must be able to apply it. The same thing is true about Professor Haig's [Hague's in original] theory. It is also not practicable, because, what are you going to do in a case like this? You buy stock, pay 100,000 dollars; at the end of the year it is worth 150,000 dollars, two days afterwards it is worth only 50,000 dollars. Now, government taxes the income, government does not pay your losses. In the first of the year, the stock has gone up 50,000 dollars; you must pay taxes on these 50,000. Next year you lose 100,000 dollars; you get nothing. Is that practicable? Of course not.

When the Supreme Court had before it the question whether or not stock dividends should be taxed, I worked out the theory that appreciation may be income whether actually realized or not, but you need realization, you must have actual realization. If you actually sell and get the amount, there is no reason why you should not call it income, except that it is going to get you into trouble with losses. When losses are made in the same year, you can deduct from one year to the other. England takes Fisher's view to a certain extent. We, in our income tax law, count capital gains as income but it works so badly that we attempt to compromise. In the present law, we only take 12%. There is some justification for England having their system, there is no justification for our law. The whole tendency in income taxation is due to a pretty sharp distinction between capital gains and income but we must take a broader view of income. The Supreme Court adapted my views, and the decision was to make realization the test. I simply call your attention of the difficulty. The scientists have not agreed to find out what is income.

The second criterion which is going to be emphasized is that of realization and separation (realization being the money's worth). The main reason why England has held back is that in the United States the ups and downs are much more common. England being an older country, the conditions are more stable and regular. In Germany the reaction has been very great, because the other system was adapted for compromise with the communists. They wanted to confiscate the property, but the government made a compromise; they set as income the capital income concept. Germany has given this up, and we have practically given it up.

If you want to have a definition of income, about as close as we get will be this:

> Income is that which comes in within a definite period in money or easily calculable money's worth in a realized and separate form above all necessary expenses of acquisition and which is available for one's own disposition.

The other part of the definition, which used to be given, is no longer accepted: "as a regular and permanent activity". This is omitted. All this time has been spent to define this. But suppose we have a very adequate definition. Even then we have some difficulty, because a second weakness involves the amount and character of income. We will seek to remedy these points. We have introduced a graduated scale for earned or unearned income. Graduation and definition of taxation will be discussed in the following two or three lectures.

In the third place, the difficulty of income as a norm of taxation is its inadequacy. Consider the case of a bachelor and a man having many children, though there are a great many other things besides children. Take two people with the same income, 10,000 dollars a year each. One man is a bachelor, stingy, does not spend, lives in a little town, pays little rent, is healthy and does not pay anything to the dentist or the doctor. On the other hand, the other man has 10 to 12 children, bad health, lives in the city, pays high rent, and is generous, etc. Now, these two men haven't the same capacity to pay.

The fourth inadequacy to be considered is the social factor, socially desirable things. One man spends all his income; another takes out a life insurance policy. Take savings, which do not come automatically but with great difficulty. We must take all these points into consideration, such that income is not theoretically an ideal norm of taxation, although it is the best we have so far. A second point under this last heading is the difficulty connected between the amount and character of the income. Legislation has gotten only to this point and it took about 100 years to get here.

## 51. Graduation of Taxation, or Progressive Taxation

Discussion nowadays is on the Progressive side. We will first take up the history of progressive taxation.

## 52. History of Progressive Taxation

There are three great stages in modern times:

(1) The Mediaeval Stage. In Italian towns during the 15th, 16th and 17th centuries, there was a great struggle between democracy and feudalism. In his book, Guicciardini tells of debates that took place in Mediaeval times during which you do not find anything that is not said even today in Washington. This was in Florence at that time.
(2) The period of initiation. This was during the French Revolution in 1792, where again people had to pay taxes according to their faculty. That was

perhaps the thing that brought down upon the French the wrath of England, that caused all the trouble in England because of the communistic character of these systems.

(3) The modern movement of graduated taxation. This is so interesting that it is divided into several stages: (a) The prelude, beginning in 1850 and lasting for 20 years. England by that time had introduced the income tax and slowly after that they introduced the system which marked the beginnings of a departure from a proportional system. In the Civil War in this country, we introduced progressive taxation. (b) The beginnings of the modern movement in 1870 to 1890. Where two great democracies began to adopt progression – in Switzerland in income or property taxes and in Australia applied to the land and inheritance taxes. Then we come to (c) The lateral spread, 1890 to 1909. By that time the United States began to apply the system to the inheritance tax. England, on the other hand, also was beginning to apply progression to the income tax. (d) You have the desired extension of the system which lasted from 1909 to 1914, in the income tax in both England and the United States. (e) The enormous increase due to the war of 1914 and up to the present time.

## 53. **Facts of Progressive Taxation**

The facts as we find them today and the last few years in United States.

We began with the income tax in 1913; then in 1916, the surtax ran up to 13%. In 1917, the normal tax was 4%, and the surtax ran up to 50%. In 1918, the normal tax went up to 12%, and the surtax went to 65%. In 1920, the normal tax was reduced to 8%, and the surtax remained the same, making a total of 73%. In 1922 the surtax was reduced to 50%, in 1924 to 40%, and in 1926 to 20%. So that today the normal rates are from $1\frac{1}{2}$% to 5% and the surtax is 20%.

In Great Britain the rates were 1.3 before the war and $6\frac{1}{3}$ supertax, making the total about 12% [sic]. Then in November the rates were doubled. We may say the high water mark was reached in 1920. In 1920 the normal tax reached 30% and the supertax was the same, making a total of 12 shillings per pound, which is 60%. During the last two or three years, they remained at 20% and the surtax at 30%. So that the highest rate is 50%. In Canada the highest rates went up to 25%. In France the income tax has gone up from year to year. It is different from England and the United States on account of the difficulty of war financing and in the last year or two the rate of tax in France is about 60%. In

Germany also rates went up to 60%. In the last few years the maximum rate has gone up to 40%.

As to the inheritance tax in the United States, we started with rates from 1 to 10%. In 1912 it increased to 20%, in 1919 to 25%, and in 1924 reached to 40%. Then reduction came and it was reduced to 20%. In Great Britain graduation came much earlier. In 1894 it was 8%; in 1901, under the influence of the radical movement, it ran up to 15%; and before the war to 20%; and in 1919, it increased to 40%, which is now the situation. In France the rates are very much higher. They have a tax both on the whole estate and on a share. The tax on the estate after the law of 1920 went up to 39% and the tax on share to 59%, so that the man who received the inheritance would pay 90% [sic]. However, the law is that it should not go over 80%. In Germany in 1919 it was 70%; it is now 60%. In the United States different states have different rates; they run to 8, 10 and 12%, and two or three years ago they reached to 15%. In California, Oklahoma and New Jersey it reached to 16%, which is added to the Federal tax. The rate on collateral ran very much higher, and reached to 40%. At the present time we find 20% in Arkansas, 25% in West Virginia, and 30% in Wisconsin. Those are the chief examples today of income and inheritance taxes.

In the property tax in Germany in 1919 the rate went up to 65%. In Italy, following the law of 1920, it was 50%. Perhaps the greatest examples of progression will be found in war taxes, excess profit taxes, of which we shall speak later. In this country, the tax rate was 65% in 1918, but that took place only when the excess profit was 40%. In Great Britain it started with no progression and ran up at one time to over 60%, but taken when excess profits were 30% of the capital. In France the rates went up to 80%. In Germany also 80% in 1919. In Italy it went up to 60%. The property increment tax, on an increase in a man's wealth, in Germany went to 100%. If a man made 90,000 dollars all the estate went to the Government. In Italy 80% was taken away. When we speak of progressive taxation we have to deal with rates which compare with early rates. [Much ambiguity here.]

Now we come to the methods and consider a subject which is only beginning to attract attention. Progression implies a relation between the different stages or records or installments and the rate of increase in the successive stages, or installments. There are two great classes of examples: e.g. where the stages are even, for instance, if we have $100, $1000 income, and where the stages are uneven. Devoting ourselves first to the question of those examples where the stages are even, we find that there are two main classes of progressive taxation. When you are dealing with even jumps we have: 1. Uniform or regular progression, and 2. Irregular progression.

1. Uniform or regular progression. There are no less than three different methods of this progression: (a) We can apply to each stage a rate of increase which remains the same (even increase) as for instance:

| | Rate of Increase | Tax |
|---|---|---|
| 1,000 | | 1.0% |
| 1,100 | 1 mil | 1.1% |
| 1,200 | 1 mil | 1.2% |
| 1,300 | 1 mil | 1.3% & etc. |

But (b) this rate of increase may grow uniformly or evenly; then we have a system which might be called a system of accelerated progression. But the rate might suffer a diminution and (c) is the method of proportionally retarded progression.

Then, progression might be irregular, which also is done by means of three methods.

| | Rate of Increase | Tax |
|---|---|---|
| 1,000 | | 1.0% |
| 1,100 | 1 mil | 1.1% |
| 1,200 | 3 mils | 1.4% |
| 1,300 | 2 mils | 1.6% |

(b) Rate of progression accelerated irregularly

| | | |
|---|---|---|
| 1,000 | | 1.0% |
| 1,100 | 1 mil | 1.1% |
| 1,200 | 4 mils | 1.5% |
| 1,300 | 9 mils | 2.4% |

(c) Method of irregularly retarded rate

| | | |
|---|---|---|
| 1,000 | | 1.0% |
| 1,100 | 1 mil | 1.1% |
| 1,200 | 5 mils | 1.6% |
| 1,300 | 3 mils | 1.9% & etc. |

All those methods are found arithmetically. All these apply where the stages have been even; then you come to infinity, where the stages are uneven. There are so many examples that it is very difficult to explain all.

Brauer [?] published a book entitled *History of Tariff* [?].[176] He gives about twelve criteria of classification. But perhaps we can get a clearer picture between our tax in this country and those of France, England and Germany by having a simple classification which would be clearer. I would classify into two great classes: Degression and Progression. Taking the system of degression, a good deal of which is found in the world, we have following important systems:

(1) The exemption of the minimum of subsistence. As in the case of allowing so much for married people.
(2) The system of abatement of fixed amounts. You pay 3% and then 4 to 5%.
(3) Reduction of the rates or abatements of percentages.
(4) Fractional assessments of the taxable unit.
(5) Where you have deductions for dependents. In this country, deductions for children; in other countries different classes of relatives.

These methods apply to degressive taxation, where everything is below normal cost. Now we come to the more important part of the question – progressive rates, increasing the rates for people with large income. We have at the present time four chief methods of progression:

(1) Where the rates increase with a total unit of the base.
(2) Much more common is the system now introduced where the rates increase with each successive bracket or fraction. So that, when we say that the total rate in this country is 77%, that is not quite exact, as that rate applies only to the final bracket. But on account of this principle, when the proper rate was applied, you will find that maximum was a little less and instead of say 65%, it is 61%. We have that system in Great Britain, the United States, Germany and Italy.
(3) You may have the fractional assessment of each unit of the base. Take the French income tax as an example. Between 6,000 and 20,000 francs today a man is assessed only at 1/25th of the rate; between 20 and 30 thousand, 2/25th; and then so on up to 25/35 until you get to 550,000 francs.
(4) We have the methods of fixing the taxable base by a progressive capitalization of the income. The best examples are Australia and Italy. In Australia the following is the formula:

$$R = \left(3 + \frac{I}{181.058}\right) d$$

where R = rate; I = taxable income [indecipherable]

In Italy Y is the rate, X is the amount of capital or income, C is a logarithmic fraction, then

$$Y = 0.04187 \times 0.438937.$$

So that there are as many records as items of progressive taxation as there are kinds of taxation. Nothing as yet is discovered to exactly suit the whole problem. We dealt with the history, we now come to the theory. (The facts also have been dealt with).

## 54. **Theory of Progressive Taxation**

Why should we have progressive taxation? Is it economically wise? There are three main definitions [positions?]:

(1) The socialist theory is that, as private property is wrong, therefore any method by which you take property away would be wise. That was the theory of the French Revolution and down to middle of the 19th century.
(2) About the middle of the 19th century a young woman, who won a prize, came out with a compensatory theory. President Walker came independently to this theory. In the case of a law, etc. making it a little easier for a rich man than the poor man, when it comes to taking away, take away from the rich a little more than from the poor man, that is the compensatory theory. This is a beautiful theory, if you could prove that two wrongs make a right; on the other hand, the one inequality is only intensified by the other.
(3) The economic theory of progressive taxation dates back to the beginnings of the discussion about benefits and about ability to pay. In the older theory you find few people advocating progressive taxation because they say the benefits of government grow faster than the theory [sic]. Nowadays, the whole theory is based upon the faculty theory. John Stuart Mill limited it to sacrifice, equality of sacrifice.

The early side involving the sacrifice doctrine was that the same proportion taken from the income of a rich man as from a poor man, fell very much more heavily upon the poor man. In one case you taxed the luxury; in the other case you taxed the necessity. Rousseau emphasized this in the 18th century. John Stuart Mill's thought was not exact enough.

In modern times the consumption side has been supplemented by the production side. As Adam Smith said, the first 1000 pounds are the hardest to acquire. If you add the production to the consumption side of faculty and the ability theory, you get, on the whole, a fairly good defense which responds to the general feeling of the average man today. What was limited a century ago to a few individuals now is widespread. In the long run you are going to get a little closer regarding equality. To work out a scheme, a plan, which would be ideal or which would be a mathematically acceptable unique method, has failed. They all depend upon certain assumptions which are arbitrary. The trouble is that when you are applying a mathematical formula you are attempting to do the impossible, because when you deal with sacrifice you are dealing with an individual matter. I may be giving the same amount, but the amount of sacrifice might be different. You cannot reduce individual psychology to a social group. All that you can hope to accomplish is to say, taking the classes as a group and in the long run[,] you get a little closer to the real sacrifice by a doctrine of some kind of a progressive scale. Whether that means a maximum rate of 70%, 90% or more or less, no one can tell. Nowadays we have much fewer objections to graduated taxation. Bastable, in his book, is still a great opponent of graduated taxation. He tells of objections as being: uncertainty; doctrine of confiscation; fine industry; unproductiveness; many other evils. Progressive taxation is an acquisition of modern democracy. Acquisition of democracy came in Florence but did not last. We now come to the other phase. We set two difficulties, one connected with the size of income, the other difficulty connected with the kind of income.

## 55. **Differentiation of Taxation**

Differentiation means difference. Progression also is differentiation, but it means amount. By this we mean differentiation in kind. The Italians call it diversification. There are really four points involved. You can differentiate taxes as follows.

(1) According to the origin and source of the income
(2) You can discuss the conditions of the wealth
(3) You can discuss the disposition of the wealth
(4) You can discuss the character of the tax payer

There are four different problems and each one comprises quite a number of individual problems. First is that which is really discussed in the books, namely, according to the origin of the source of the wealth. That you find in the English literature and the names are very numerous. We have discussed the distinction between funded and unfunded income. Sometimes it is called the distinction between permanent and perpetual income. Sometimes it is called the distinction between life and precarious income. Sometimes it is called the distinction between inheritable and terminable income. Sometimes it is called the distinction between property and labor income. Glasgow made a distinction between what you call lazy and active income. The term that we use is the common distinction, that between earned and unearned income, sometimes called investment income. Now, the distinction really began with the introduction of the income tax itself in England in 1780. It was not until after the year 1850 that active members of Parliament took the matter up and got the support of John Stuart Mill. However, this failed because Gladstone was in opposition. It was not until this century that the radical movement became so important in England and the special Royal Committee was appointed to consider the question in 1905–1906. So that in England it was adopted in 1907, and since that time every year in England they have made the distinction, and in 1920 there was a 10% off for the tax on earned income.

In the United States although recommended by me, yet Mr. Hull[177] said very much what Turgot said as regards abolishing feudal privileges and at the same time abolishing it for the clergy. He said one at a time. For the following ten years the objection was always made, it was found too complicated. In 1924 they introduced it and it worked "like a charm". In 1924 we introduced the system of taking 25% off for earned labor incomes but limited it to 10,000 dollars. There was no logical defense of that. In 1926, we increased it to [20,000] dollars. There is of course a certain theoretical justification in putting a limit on differentiation of earned income, because after a limit has been reached the production side becomes rather unimportant compared to the consumption side. It is really only in the case of small incomes that we can contrast the earned and unearned income. Other countries have gone further. The difference between earned and unearned income is more or less arbitrary. You buy a farm, is your income earned or unearned? You have your own farm and hire farmers to do your work for you, is your income [investment in original] earned or unearned? As soon as you ask these questions you get into difficulty. Italy was the first country to attempt to answer this question more sensibly and instead of making a differentiation into earned and unearned they have divided income into five or six classes. The highest tax was imposed upon government bonds and securities, a little lower rate on income from houses, a still lower rate for agricultural land, a little lower rate for professional men and wages, and lowest of all government salaries.

In the second stage, we have difference according to the nature or condition of the wealth. Differentiation of the differences that we find in the world today under this head are in very great classes. (a) Differences due to public policy. (b) Differences due to administrative convenience. The forms of income tax. The income tax in England is divided into schedules. There are five different criteria: income, rent, business, etc. And yet in England it is mainly an administrative device and the rates are almost all the time differ [determined?] from one or two of the schedules.

In France, however, they have a different tax, certainly a different rate upon each schedule, and other countries often follow the same rule. We apply the same rate about everything. In Massachusetts they distinguish between income from salaries and have different rates, so that you do find examples of differentiation in the income tax. When we come to the benefit tax, we come to something which is very peculiarly American. Under our system of property taxation, the general and special property tax, we now in this country frequently employ a differentiation. We do it not alone in our state and local taxes, but also even in some of our federal business taxes. The question arose before the courts as to whether that was legitimate so far as the federal government is concerned. We call that whole subject in this country classification of taxation.

When we come to our state or local taxes we find there is a great fight over that. We shall see [that] the general property tax has been breaking down in this country and we shall also see that among other attempts there are two attempts to get rid of the evils. (a) By classifying property taxes. (b) By getting rid of the whole system and introducing an income tax. A good many states which are not yet ready [for an income tax] attempt to get around the evils of a so-called uniform tax by classifying property. This classification is in reality not a new thing at all; we find it in the history of taxation both in this country and everywhere else. Taxation of property always becomes a tax upon land; {then you had this kind of property as they come along you get general property tax and when you reach the apex declined and classify property and we get rid of it.} You find this condition in the 18th century in Vermont, different rates for land, horses, and property.

When we speak of classification, the classified property tax in this country today, we mean by classification that they refer to the modern development of taking out certain kinds of property. Certain others almost evade taxation, putting a much lower rate on securities, mortgages and so forth. We find today 12 to 15 states that follow this classification. Those who are interested in facts will find the very best facts in the Annual Reports of our New York State Taxation [sic]. They now give about 10 to 20 pages of reports with reference to the facts of every state in the country. That is the chief form of differentiation which is in active politics in this country today. The classified tax, or classification of the property tax, is an immense improvement on the general property tax but even this does not go far enough and in this state we have abolished all [of] this system. The second form of differentiation is not given in the books at all, and it is a very important one in this country.

The third differentiation is according to the disposition of [all?] wealth. Under this heading there are two problems: What you save and what you spend. The theoretical defence of exempting savings is very admirably put forward by an Italian in this country. It was suggested by T. S. Adams,[178] but has not gone into legislation. The distinction in this respect still is not made. The other is, what do you do with your income? When you give it away, we allow deduction of all gifts provided it does not exceed 15%. We have limited this, however, for associations and corporations, because gifts may go to private individuals. At all events, you see, there is the beginning of a system which refers to what you do with your wealth.

The fourth differentiation is done according to the character of the taxpayer. We already make a distinction according to whether a man is a foreigner or native, whether he is a resident or nonresident. Some countries make a distinction whether a man is married or not, and that point comes up under the head of double taxation.

## 56. II. **Universality of Taxation**

We will now discuss the universality of taxation, which is also a product of modern times – one of the forms of application of modern principles, principles that not only should all be treated equally but that everyone should be taxed and should bear the burden. Now, in the matter of universality, again, there are two aspects:

(1) Everybody should be taxed
(2) Everybody should be taxed only once, not one man taxed once and another man taxed twice – higher principles of justice. What we will take up now under this first heading is the problem of exemptions. To what extent is that legitimate under the rule of universality?

## 57. **Exemptions**

We know it nowadays, exemption or immunity from taxation, and this, as you remember from our discussion of history, was the original term in the Roman Empire – when it had reference to freedom from those very burdensome charges, known as munera, and charges of all kinds. In those days, the freedom was called immunity. Throughout the Middle Ages that term was applied. These exemptions were originally found in two cases: (a) Nobility, and (b) Clergy, and in each case there was a certain historical justification. The noble man was exempt from payment because he had to serve in person in the feudal array in the war; he also had to provide his horses and other equipment. He was not subject to these taxes which others had to pay, that was a perfectly fair proposition. So, also, the clergy often received exemption from taxation, because there was dedicated to the clergy certain duties which were not left to the state, namely, the support of the poor, seemingly an exemption for what in reality if not done through the clergy would have been charged to the government.

The original system of exemption soon became so abused by the nobility that when the feudal array disappeared, after the invention of gun powder, this turned into a privilege which was really undeserved. This freedom now extended to all other kinds of taxes, so that the nobility was virtually free from most of the burden. With the clergy, too, even before the Reformation, a great many duties, which originally involved the church, were gradually transferred to the state, and the clergy now had more wide-reaching privileges. Not before very long the whole mass of taxation began to press not only upon the laymen but upon the poorer classes. Because not only the nobility and clergy were exempt but also after a while practically nobody who could pay for the privileges – like judges, the legal profession – paid taxes; gradually exemption entered into whole other professions. So that, shortly before the Reformation, you had a system of taxation which was honeycombed with these various class exemptions.

What was found in France was also found in other countries on the continent. In England, the nobility were fundamentally important, and the situation as regards to the clergy had changed. The English nobility now began to assume the duties and responsibilities of government; they undertook the work of the government and subjected themselves to all sorts of burdens. In France these people were known as the privileged classes, and efforts at fiscal reform began in the 16th century and ran all the way down to the Revolution, the purpose being to diminish if possible the spread of these privileges. Turgot himself made a good beginning but was overthrown by *gabelle*.[179] It was not until the Revolution that the feudal privileges disappeared and the principle of universality was laid down in the constitution and spread all over the continent. So that all systems of exemption and privileges disappeared and in its place developed a new system of exemptions. Nowadays, we find that this can be put under five heads, exemptions due to one of five reasons:

(1) Diplomatic exemptions
(2) The exemption connected with the idea of ability or faculty to pay
(3) The exemption connected with public policy
(4) The exemption connected with public credit
(5) The exemptions of government purpose itself

We will say a few words about each kind of these exemptions:

The first calls for only mere mention: That in a country a foreigner is exempted from taxation, from prohibition laws, and practically from every kind of domestic laws, except the laws dealing with crime and such. The diplomatic exemption is after all of very little importance.

Quite different, however, is the second class. The exemption is connected with the idea of ability or faculty to pay, and originated with the conception of the minimum of subsistence. People who cannot afford to pay taxes because they are at the very minimum of subsistence should not be required to pay taxes. We find this idea at the beginning of the 18th century, the benefit idea [sic], and others have been especially developed in the 19th century and also at the present time. There are two problems: the first is, what is the actual minimum of subsistence?, and second, what are the arguments for and against? The argument in favor of exempting the minimum of subsistence is the obvious one, that if a man is very poor you cannot get anything out of him; and if you make him pay, that the community has to pay him back by giving him support in order enable him to live. If it is the actual minimum, the argument advanced in refutation of that view is that it is better to do the latter than the former, better to make everybody pay even those the government has to support, otherwise democracy would bring dangerous conditions between those you have to support and those who can pay. The arguments are that if you exempt rich people, who nevertheless can afford and decide upon their expenditure, you necessarily introduce extravagance which is dangerous even to the maintenance to the state. (A very interesting book is written by a German on the minimum of subsistence, which was first put forth by Professor Cohn and others. He points out some explanations of undue principle in Austria and other countries.) The answer to that is that under our American system of the 19th century general property tax, you had this system embossed more with a minimum of subsistence at that time in this country. A man who spent all he made never was taxed because he did not accumulate anything, and we have never found any great danger arising from that; even if we did it would be useless to change. It is very easy to give but difficult to take. This argument is not very great. The only point is that if you unduly raise this limit of exemption, then, of course, you may have, and we have had, some difficulty connected with it, but in general the principle would be perfectly sound.

What is the minimum of subsistence is the other side of the doctrine. At first it was a bare minimum of subsistence, but this standard changed, because in the past the minimum was put too low. Then the claim was put forth that by a minimum of subsistence was meant a comfortable minimum of subsistence, not only to give a man comfortable support but to enable him to send a child or two to college. As far as that goes the minimum of subsistence nowadays should include a radio or an auto. There is, of course, no limit to this and if stretched a little further the minimum becomes absurd, but when you once bring wages up it was very difficult indeed to reduce. When we had our income tax reduced to the minimum, to the point to which it was reduced, our minimum income tax was 1,500 dollars, but married men were allowed 3,500 dollars, and the minimum in the original law was of course not a [subsistence] minimum, it was a political minimum. That law was passed by one of the great parties, the Democratic Party. It came from the South and in order to be able to free most of the farmers they fixed it at $4,000, which practically freed all the farming class.

When you are dealing with a political problem you get away from economic principles. In other countries, of course, this is much less because the standard of living is much lower. It may be said that you find everywhere exemption of a minimum which is little more than a bare minimum of existence. In England they only exempted 100 pounds, but in this country the exemption goes up very much higher, some up to $50,000, some $75,000 and some $100,000. In these, of course, we cannot speak of minimum of subsistence. In our state and local taxation in this country you find exemption upon the small property tax. You also find in some of our states, and for that matter in other countries, a general rule exempting individuals from taxation because of a presumed lack of ability that goes beyond minimum of subsistence. You find this in the tax laws of Massachusetts and Alabama. Tax authorities have the right to reduce whenever there is reason to believe that there is general loss of ability, but in the main this doctrine of minimum of subsistence is accepted even though with certain limitations.

The third exemption is in connection with Public Policy; these are very common and widespread. We take this country. In connection with our property tax, growing corporations, exempting mechanics' tools, teachers and farmers, then more recently soldiers and sailors, especially veterans of the war. Savings banks, charitable organizations in general and cemeteries. Then a step forward brings us to manufactures, industry, the same principle that was responsible for the adoption by the nation of the doctrine of protection to industry. The same idea is responsible for the exemption either temporarily or more permanently of certain classes of industries. When in this state as well as in Pennsylvania certain forms of taxes in business were adopted, the manufacturing industries were freed. Some states freed only for a few years, such as Louisiana and Vermont, and in other places they freed in general. At the present time they always run against a stone wall in Pennsylvania. In this state this was abolished a generation ago when we had reform of taxation. The legitimacy of such exemption can only be temporary and not permanent. The same thing is found in the case of houses in the city; they were exempt for 5 years. Now, again, we have in Albany this discussion, exempting houses where dividends are limited. In regard to certain other exemptions, such as schools, educational and philanthropic institutions and ministers, there is no debate at all. This is also true with certain kinds of banks.

When it comes to churches we have great trouble with that. One difficulty we have gotten into is one of our problems of exemptions. The churches originally were exempt, very much for the same idea that was responsible for the exemption of clergy in the Middle Ages. Now, we have separated the church from the state. Church duties and dues were at one time considered a political duty. When the Catholic church began to get a great deal of property about a half century ago and went to the community which was Protestant, the New England and Southern states went into the flame of sectarianism. We had a great movement of exemption of religious property, but it passed away after a time, although we hear of it now and again. The matter came to prominence about a generation ago in this city when the worst tenement house was owned by the Trinity Group which had exemption from taxation. That created a great deal of comment at that time, and the disappearance of taxation by the church itself was realized. On the whole, this ecclesiastical exemption is thought to be a legitimate and defensible one. This kind of exemption will probably continue.

In former times we also had an exemption of railroads which was a source of great trouble, as those were granted exemptions before the 40's. Daniel Webster's college exemption also was a lasting one, which was considered against the constitution and since the beginning of the 40's irredeemable exemptions have been rendered impossible. That also is a matter of minor importance.

The fourth is the exemption connected with public credit. This actually was one of very great difficulty. It is connected with the exemption of government bonds. There are two problems under this head in this country: (1) whether government should exempt its bonds from taxation and (2) the question of the relation of state and federal taxation: the exemption of one form of government from taxation by another form. Taking the first question, is it or is it not wise to exempt its bonds from taxation? These bonds were originally issued in times of great emergency – time of war – a time of great difficulty when the credit of government suffers. Since taxation becomes almost impossible, in order to save itself from the issue of irredeemable paper money, the government marketed its bonds by making a great inducement, saying if you buy these bonds you will be exempted from taxation. It is all right when the war breaks out, then what happens to the bond after the normal conditions return? Here, of course, you have a condition where the whole class of bondholders have a right to receive money from the government but are not called upon to pay anything to the government and that creates difficulty. The difficulty of injustice is not as great as it seems, because the exemption at the time when the bonds are marketed is capitalized into a higher price paid for the issue of government bonds of 6%. When war breaks out, you know you are going to be exempt during the life of the bond, and you are willing to pay somewhat more, so that [payment?] is already made in the beginning. But that is not true where you have a system of graduated taxation, because in a system of graduated taxation it is only at the point of the margin that the capitalization takes place and people subjected to higher rates of graduated tax would enjoy an unearned, unlooked for preference. The real decision depends upon the acuteness of political exigency, the maintenance of public credit, and the maintenance of justice and equality in different classes of people. In general, so far as it goes we ought to get away from it.

If the country is in a very great difficulty, if this seems absolutely necessary, you have other considerations. Nine times out of ten if you tempt capitalists, they will come forth with their money, so that on general principles the policy should be to limit the exemption to the lowest possible degree. In 1917 some people pleaded with the Secretary of the Treasury when we were putting out bonds, which has been so long a custom in this country, that it was difficult to bring about any change. In England they are experienced and they issued some bonds with exemption and some without exemption.

What we have said about national bonds and securities applies to local and state securities. During the last decade or two we issued local and state securities with this exemption. The argument at first was a good one. It took issue with state bonds for schools and roads. The exemption would enable them to issue those bonds at a lower rate of interest. Instead of paying $4\frac{1}{2}$%, pay $3\frac{1}{2}$%, which meant a lot of saving to the taxpayer. This is a strong argument, but this advantage disappears if a state begins to issue tax-exempt securities over against securities which are taxable in proportion, and the state issues more and more securities. The Secretary of the Treasury in Washington pointed this out even in 1918–1919. He said that it is true that it gradually weakens and the advantage disappears as the amount of issue becomes greater. Where you no longer have this narrow field of tax-exempt bonds over against the immense field of taxable bonds, when the field becomes more and more taxable [sic] the advantage disappears. You can prove this from year to year according to the issue. As you get closer and closer to the saturation point the advantage disappears. So the argument in favor of non-exemption is among the strongest.

The chief difficulty comes from the second stage, owing to complications between the state and Federal governments. We have the very great additional difficulty under our constitutional interpretation, as federal bonds are exempt from state and local taxation and state and local bonds are exempt from Federal taxation. It is true that the second situation is a result of misinterpretation of the economic cases by the judges. That federal securities should be exempt was due to Marshall,[180] based upon a sound economic decision, but the second, namely that state and local securities are exempt from federal taxation, was based upon the alleged same condition, which does not exist.

This was on account of the failure by the judges to understand the real situation. It is now difficult to do away with this and this is due entirely to political, not economic, considerations. This is in reference to the decision of the Supreme Court. The practical result is, especially in the last few years, since state and local, especially municipal, debt increased so enormously from $1\frac{1}{2}$ to 2 billions a year, that we are adding this amount; already we have 12 to 15 billions outstanding. If this goes on we shall have a very large part of all the securities in the country, public or private, consist of nontaxables. So that it will be still easier than it is today for an individual to invest in nontaxables to escape from taxation entirely. This led Secretary Mellon to bring about a change by offering a change to the states. He said to the states, if you will abandon your privilege of exemption for your bonds, we will abandon the privilege from our bonds. That almost went through, but it failed because of a resuscitation of the idea of states' rights, of state sovereignty, which was unwarranted. In the constitution, of course, it is expressed and implied that the federal government will do nothing against the equality of the states. There is nothing, however, in the constitution that will give the states unwarranted privilege. The situation, therefore, so far as the states in this country are concerned, is very confused. When you come to the 16th Amendment it goes the other way. As for retention of the state tax by the states, you find the greatest advocate of centralization in our Southern states, so the whole thing is confused today.

In some respects there are arguments in favor and in other respects it is the other way. But from the point of view of science and theory, there is no doubt in both of these cases. Especially in the second case, namely, state versus Federal exemption, it would be wiser to abolish that exemption altogether, whereas in even the first case the argument is strong. The conclusion is therefore that the whole tendency of increasing the range of exemptions is unwise.

This will bring us to the final exemption: The exemption of government on its own property. From one point of view this is of very little consequence, because it is a bookkeeping device only. State government, for instance, would tax property, would impose a tax upon capital, school houses and other things, then the tax will have to be paid, of course, by the state and taken out of revenues; whereas if the state exempted that property, the situation would be precisely the same. This, therefore, is simply a bookkeeping device. Abroad they generally tax their own property. The real difficulty has arisen in this country, again, because of inter-state, local-state and federal taxation. The question is whether federal property – in the case of a school house, fire houses, or courts – and state property are subject to local taxation and whether the local school house is subject to state taxation. There was a question of state hospitals in this state; we built those hospitals in a local community and the amount of the real estate amounted to a considerable proportion of the total tax roll of that community. This meant a great increase in expenses for hospitals, and the question arose, why should we have these hospitals bear all the burdens? So the situation of so-called property of some other jurisdiction has become very acute in this community where you have state property and federal property. In order to bring about the realization of equality of burden you will find that the amount of exemption in real estate has become enormous, aggregating to a very large sum; and everyone feels that if we get rid of this exemption the taxation would go down indeed. For this particular system it would be better to introduce the European system and have each form of government pay taxes on its own property. There is perhaps one exception, which exempts [increases in the original] the educational, philanthropic and even religious institutions. The arguments, however, seem in modern times to be in favor of restriction rather than extension of exemptions. But the problem will still be with us for many years in different ways, and will, in all events, have this satisfaction, that exemption in modern times is no longer open to those very grave objections which attached to the problem in mediaeval and ancient times.

## 58. Double Taxation

This is the problem of taxing one man once and another man twice. I will sum up in a few words the more recent developments of the last few years. By double taxation we mean the unjust double taxation. Various taxation does not mean double taxation. By double taxation we mean the unjust or unequal taxation where one man has got to pay more than the other; it is the injustice or the inequality that we are dealing with. You remember, from the *Essays*,[181] that there are two main categories:

## 59. By the same jurisdiction

## 60. By competing jurisdictions

A man in the first place may be taxed upon his property and another man may be taxed upon his income as well as his property, one man may have income and another may have income and property. Taxation of income or property sometimes is the only device for securing the differentiation of taxation, such as in Russia before the war and even today. What should we do with debts? One man buys a farm and pays $10,000, another man borrows $5,000 to buy his farm. What shall we do with the mortgage? That whole problem we will dismiss at present with the remark that from a social and economic point of view, there are certain better methods for dealing with mortgages.

The next is the interesting part of the problem, and that is, what are we going to do with the dividends of the corporation so far as the individual is concerned? You may tax the corporation and exempt the dividends of the stockholder, or you may tax the corporation and not exempt the dividends of stockholders from the tax, as we do in New York State. Or you may adopt a compromise, where we have the corporation income tax, where dividends are exempt from normal tax but not exempt from super tax. There are problems, not many, that still exist so far as taxation by the same authorities is concerned.

Now, then, we come to the other side of the question, e.g. simultaneous taxation by competing authorities. To bring the situation home to you, let me give you one example, a case that occurred here a few years ago: Mr. U. was a citizen of state A, but happened to die in state B. He kept his securities over across the river in Jersey City, state C. He had securities, railroad bonds, mortgage, etc., the headquarters of which were in Chicago, state D. Each of the four states claimed taxes upon the estate of this man. This explains what is apt to happen. Another example will be the case of a corporation doing business in China and Malay, or New York. Its [His in original] entire income is subject to American income tax, and China may impose a similar tax. In the case of Navigation companies, big steamers go all over the world. It is such cases that have made the problem so acute. Astor[182] is another example, who gave up his American citizenship during the war in order to get the title of Baron in England. He had to pay income tax here from his income received and for local tax also. He spent 120%. Some people will say that it serves him right, but we do not look at this from this standpoint.

Back in the Middle Ages you will find the first discussion, which didn't amount to very much, as to whether a community could tax a foreigner. At that time the problem was easily solved because of political allegiance. People coming from other towns in the same country were considered foreigners. There was no difficulty until people had international economic dealings and capitalists owning property in foreign lands. Gradually the problem shifted after a while. It was customary, so far as land was concerned, to be taxed without respect for who owns it and for where it is situated. In other things they said he ought to be taxed wherever he resides. This is not applicable at this time, as you know where land is but you do not know where the origin of income is. Take this case: A tea company has headquarters in Holland, the stockholders may live in England, and the income of that tea company comes from the business consisting of a tea plantation in Malay. Then, of course, tea is going to be sent to France, and is finally consumed by somebody in Germany; and it is not until the last bill is paid that dividends accrue to the tea company. Where is the origin of that income? Is it France, Holland, Germany, or England? This shows the complications that can rise. Even as regards property you get into difficulty. For instance, sheep may be in one state one day and go from one place to [an]other place. This has entered into politics. There have been three kinds of attempts to remedy the situation: An attempt was made by the separate states in this country a generation ago, and in Germany also. They got around the difficulty by a law empowering the Federal Government to make the decision that land should be taxed by the state wherever it is located, where the fixed business is, etc. In Switzerland, they had a great difficulty in the 70s. They decided that it should be regulated by the separate states. In this country we have neither the one nor the other. The Federal Constitution does not give anybody any power to settle that question, but under a certain clause, the 15th Amendment, it has become rather customary now that we have a common decision of the Supreme Court of the United States which somewhat affected some phases of this question. It brought about a little improvement and still there is great trouble in the general [property] tax. The property tax itself has never been very rigorously enforced. Real trouble has arisen in three [sic] ways: (1) In connection with income taxes. Only twelve states have an income tax. (2) With reference to the abolishment of property or the income of corporations that do business in more than one state. The real step forward has been taken by members of a great empire, by the British Empire, [and] by independent states through the efforts of the League of Nations. There are three kinds of attempts to get rid of the difficulty: Attempts made by a member of a federal state, by a member of an imperial union, and by several nations.

1. Members of a Federal State. The situation arises in the United States, Germany and Switzerland. In the United States the difficulty was never great because of the breakdown of the general property tax. The difficulty arose in countries where more modern methods of taxation were introduced. In Germany they got rid of the difficulty by the interference of the Federal Government. They passed a federal law, first in 1870, then again in 1909, stating exactly what a state could do and what it could not do. This did not apply to local taxation. Now, of course, the states practically have no power.

That is one way of doing it. The next thing was in Switzerland. The problem arose in the 70's and 80's. Their constitution permitted the federal union to attend to this, but instead of passing the law they left it to the "Supreme Court" of Switzerland, elaborating all sorts of principles and settling the matter in individual cases. This is the second way. In this country you can do neither the one nor the other under our constitution. Of course, the Supreme Court has certain powers but they are limited. They can only deal within a fraction of cases, considered as to the equal protection of the law. In the Federal government, the 15th Amendment is applied to pretty much everything. We have a negative, not a positive condition. The Supreme Court can say that this particular action of a state is unconstitutional, and about the only positive results that have been attained under our system are that in the end real estate can be taxed only where it is situated and that certain kinds of personal property that can be located and are tangible can be taxed. As regards the inheritance tax, certain thing have been done by some of the states which are illegal, such as the attempt by states to tax non-residents on property outside of the state and securities which might happen to be in the state. The states are gradually beginning to do away with every one of these efforts. But the situation is still very unsatisfactory, especially where it is dependent entirely on principles of a state committee. Interstate commerce is a beautiful theory but it does not exist in this country. All we can do is to either develop more fully the powers of the Federal Supreme Court or change our system, developing public opinion in the different states leading them to make the attempt to cure the injustice of double taxation. New York, Massachusetts, and Wisconsin have already made some attempts to do away with double taxation. New York, instead of taxing corporations on all of their income, taxes upon so much of the income as is economically sound and earned within the state. But every state follows a different principle. All I can say is that within this country we are in the first stages of beginning to do away some of the evils of double taxation. Australia came during last few years.

2. The best example of this class of cases dealing with the evil is the British Empire. There the situation was taken up energetically even before the war. The situation became very important and even urgent in Australia, resulting in the appointment of a committee in 1920 after a conference in 1917. A law was passed to meet the situation. The British system is that if any person has to pay income taxes on the same income in both England and Australia, they will remit [deduct] from the English income tax levied on the man who lives in both England and Australia up to 1/2 of the amount. Other dominions also do the same thing. There you have the system of 1920, whereby a relief is afforded to

a man who lives in England, who makes his money in India or Africa. All these attempts fade away when we come to the third problem.

3. Several States. For instance, England and the United States, Germany and France, or Italy, etc. When the International Chamber of Commerce started the first thing they did was to appoint a committee to deal with the subject. The League of Nations took that up. They had four men to lay down the principles of that committee. I had the honor to serve on this committee. We organized it and after several years in Geneva it adopted a report. The report was called "The Report of the Economic Experts". Summing up the conclusions: What this report tried to do was to call attention to the economic effects of double taxation, to analyze different constituent elements of the economic allegiances of each country, and then to come to a conclusion. There are four possible methods of dealing with the problem: (a) The method of deduction, the method that we follow in this country. This has to have some restrictions, in the case of a home government at the mercy of another country, or there will be nothing left to the home country. (b) The method of exemption, which is to exempt all nonresidents from taxation, whether income, property, or anything else. This means that every nation should try to tax only its residents on their income. (c) We recognized that it would bring up great difficulty between creditor and debtor countries. In Australia, for instance, they tax people upon the income received in Australia. Australians are not taxed by other countries as there are no Australians anywhere else. But in England, the principle of taxing only residents on their entire income is, of course, a good one. That does not work in cases of investment, to which we apply the principle of division.

4. We took up the case of those taxes, which you remember from our discussion of the kinds of taxes, we called semi-personal and quasi-personal taxes. We say that in cases of that kind the fourth principle might be the principle of classification. That is, classifying in the matter of real estate, you find where the real estate is situated. {Then we took the principle of only resident.} In these other taxes perhaps another system might be added. In that particular report it so happened that every one of the members of the committee came from countries which had a modern income tax. When the report was rendered, the League of Nations invited practical administrators. The United States was not represented, not being a member of the League and therefore every member with one exception came from countries that did not have modern taxation; they came from where they had semi-personal and quasi-personal taxes. The consequence was that when the final report came in, the United States was represented by Professor Adams as a technical representative. They came to an agreement and the basis of our own report was made

somewhat more clean cut. They distinguished between what they called personal and impersonal taxation. That is the situation now. All that we can say is that great progress has been made because everybody accepts the principle in question. The United States government published a few weeks ago a report on taxation. We can say that even in the matter of the international question regarding double taxation we are now on the road to a solution, much more so than we were a few decades ago.

## 61. **The Tax System: Single versus plural taxation**

The question is, why not have a single tax? – one tax. This idea of a single tax has attracted a great many people's attention, even in the Middle Ages. In England you find a single house tax in the 18th century; you find the Physiocrats coming with their single tax on land and on real estate. Beginning with the 19th century, you find enthusiasts for the income tax, then finally Henry George came with land values, quite different from a single tax. Henry George did not know the exact situation at that time. He thought he started a new idea and had many a new convert in economics. There are, however, a few men, such as Professor Carr [Carver?], who lean that way. This movement is kept by bequests; outside of that it does not exist and there is no more single tax movement. We should not devote any more time for this. How do you explain that we do not hear of the single tax in England any more? In England it is gone. You find it in Australia and above all in some of the South American States. Now, why is that?

The general theory is that Henry George still bases private property upon the labor theory. He says that property in land is as bad as property in slaves. Slaves are not made by human beings; also land is not made by human beings. Henry George says nothing can be owned except what is made by human beings. And yet he is not a socialist. But he did not know that the labor theory of private property had been abandoned. He bases his concept of public finance upon the theory of benefits. I suggested the idea of ability to pay. You see, he is belated in both his general theory of philosophy and his theory of economics. His fiscal difficulty is that any single tax, no matter how good, is bound to intensify inequalities. Political difficulties, too, are seen; above all, as Voltaire expressed in his famous essay,[183] the economic difficulties of course are very great. There is not enough rent in a good many rural sections in this country. If you took all land values you would not be able to have enough for good schools and roads. It would require a change in our constitution and this is out of question.

How, then, does it happen that in Australia and South America you find some enthusiasts? The Australian situation is complicated. It is intelligible only because, in Australia, land is taxed upon what it yields, not upon what it is worth, whereas the single tax, like our own American land value, is laid upon value. This is a great movement from yield to capital value. The tax never has gone any further in Australia except exempting buildings from taxation. The South American situation is very different; most taxes are raised from consumption and very little is raised from land. The movement, therefore, in favor of the single tax amounts to nothing more than a trying to get taxes near to our system. So that, we pass by the whole subject, saying that no one believes nowadays in any kind of single tax, but [the existing] system[s] of taxation and those are the ones we are going to discuss from now on.

### 63. Tax Administration: Tax Commissions

In modern times tax administration is becoming fundamentally very important. In former times this did not amount to much. In the Middle Ages most taxes were frowned upon. *Fermier genereaux* became very rich. Only farmers could spend thousands of dollars [sic], farmers were the ones who could buy. "Leeches", they were called, sucking the blood out of people. Nowadays, every government collects taxes. All the problems of efficiency and economics in connection with our own income tax we have pointed out. The best kind of tax may be spoiled under bad administration. When you compare taxes you must compare administration as well as principles. It is hard for the government to debate with shrewd debaters. They may be just within the law. When we come to local administration, the situation is worse. These, of course, are the natural shortcomings of democracy. They will give Mussolini a great many points to discuss.

# BOOK III

# Taxation – Special

## Part 1. Direct Taxes: Taxes on Wealth

Classifying taxes into either personal or impersonal, applying to indirect taxes. Taxes on tobacco, whiskey, etc. The first category of these taxes will be on real estate.

## 64. **Impersonal Taxes**

## 65. I. **Tax on Real Estate**

The first problem we take up is, what do you mean by real estate? Of course, all of you know that our distinction comes from those you find in feudal times. It is a feudal distinction, very different from those you find in Roman law – multiples and non-multiples. Whereas we distinguish between real estate and personal property. Real estate with us means not simply land but also all kinds of rights attaching to land. The point is interesting because for tax purposes a great many changes have been made by legislation, in fact, not alone for tax purposes but for other purposes as well. The result is that what is designated real estate in one state is not necessarily real estate in another state and [the same is true of] personal property. For instance, take a certain thing, such as gas and water mains. Of course, ordinarily they will be called real estate. The states calling this personal property are Wisconsin, Montana, and Washington. Take railway tracts, railroad property, which is necessarily used in the operation of railways. A good many states like our own call all that property real estate; in other states, such as North Carolina, South Carolina, Iowa, etc., they call it personal property. Street railway tracts in this city are real estate, but not in South Dakota and North Dakota. New Hampshire and Arizona call the telegraph real estate, but California and South Dakota call it personal property. Wisconsin and South Carolina call bridges private property. South Carolina calls roads and turnpikes personal property. Ferries are real estate in most states, but Wisconsin says they are personal property. The franchise of incorporation that is generally called personal property, Wyoming calls real estate. The special franchise right of public utilities to use public streets and mineral rights is generally called real estate, but Arizona calls them personal property. In other words, the whole uniform feudal system has been modified. Therefore, when you compare statistics of taxes on real estate and personal property, you need to be very sure you know about such matters. The statistics can be only very slightly comparable. We are dealing with real estate, but only as the law of the state defines it; real estate does include lands and houses.

Now, having ascertained what real estate is, the next question is, how does a real estate tax fit into the general scheme of taxation? There are four different ways in which this can happen:

(1) You can tax the real estate tax as part of the general property tax.
(2) You can tax real estate alone as a tax under the impersonal tax irrespective of the individual who owns it, as in the city taxes levied against real estate – a tax *in rem*.

(3) It may be a part of your income tax with the income of real estate included in your general income tax base.
(4) It may be a part of the expenditure tax, as in France: what a man gets rather than what he owns.

Now, then, that being the real problem, we shall deal with it primarily under the first three heads, leaving the expenditure tax for a little later. We come in the next place to the problem of history, since real estate comprises three separate things: (a) land, (b) improvements on the land – houses, and (c) franchises. We shall therefore go right on to:

## 66. **Land Tax: Taxation of Forests and Mines**

The land tax is the most important in this country. This is wrestling with a very important and pretty difficult practical problem, which can be referred back to questions of principles and theory.

A few words first about the history of the land tax. It is the oldest kind of tax on property, because it is the easiest kind of tax; we find it in all the ancient civilizations of the Orient and Classical Antiquity. Later on, when religious governments began, it became very important in the form of the tithe, a 10th, that also has a certain separate important history. If we pass to the early [modern] period and attempt to explain the existing conditions, we must say a few words about the mediaeval conditions in England. In England, the first attempt to get revenue from land was by means of the so-called hideage[184] and *carentage*, which is a plough, and was based on the amount of land that could be ploughed with eight oxen and this typical plough. All these early taxes gradually merged into what in later centuries became known as the 15th and 10th. A 15th of the produce of the land was, all through the early ages, the system of the general property tax. That broke down after a while because it didn't work very well. It was again tried by Cromwell, then called monthly assessments, and that also did not work very well. After a time, in 1697, instead of a certain percentage of the produce of land, a fixed sum was paid irrespective of how much the land produced. That worked fairly well but that also gradually met with difficulty and in the year 1798 the whole mediaeval property tax was made a redeemable rent charge. It was still payable by land, by the owner of the land. One could buy land free of tax by capitalizing it. Since 1798 about half of the land was redeemed of this charge upon the land, so that today the so-called land tax does not yield more than five or six hundred thousand pounds. It is levied upon a small part of the land, the capitalization taken on a 3% basis. That is all that is left of the old land tax in England.

Now, in France the system worked out very much the same way. In England the name was different; in France they called it taille: taille reål and taille personel. It became honeycombed with abuses. It was that tax which created most of the trouble before the Revolution. The first thing they did in the Revolution was to abolish this tax. In its place they levied upon a thing – not upon the person but upon the land and called it a land tax. They decided how much should be raised for the entire state and divided it up among the departments and localities. In France the tax was levied upon what the land could produce. In order to find that out they made a survey of valuations, called the cadastre. This is simply a table, in the greatest possible detail, of the acres of land and the yield at the time. The trouble with this is that it is a very expensive thing to do, it takes a long time, and by the time it is finished it is not worth much. So in France they had a great difficulty with the survey on each particular parcel of land. As a consequence, about a half century ago they removed houses from this land tax; and since about a decade ago, instead of putting a particular obligatory amount, they now tax the land according to the yield or produce value of the land. Owing to the war, the rents went up and they have to pay more, about 10% – a tax on the thing, not on the person. This modern French system spread pretty much all over the continent. In Italy and Germany the system of taxing land is to take a certain portion of the produce of land irrespective of the land owner. That is the way the modern land tax has developed. Now, the land tax being found everywhere today, except in England, the next question is, what are the criteria of liability? How are you going to tax land? Historically there are six different methods, all of which are found today somewhere in the world:

(1) Method of taxing the quantity or amount of the land. That was the case in Rome, with the so-called *ugatio* ($2\frac{1}{2}$ acres of land) – the hideage in England. In this country, Vermont started out with such a tax. South Carolina used the tax up to 1785, North Carolina up to 1796. Japan had it. But, of course, a tax on the quantity irrespective of other considerations is an exceedingly rough kind of tax.
(2) A second system was adopted to improve upon that, by taking the gross produce of what a land yielded. They take different parts of gross produce, in India one-fourth, in Egypt one-fifth, then payable 10th, and Christian societies adopted the 10th. In the Middle Ages this tithe, of course, played a great role. There are several different kinds of tithes. It was established in the 8th century by Charlemagne. In the Middle Ages they had predial[185] tithes, or personal tithes, which applied to individual earnings, and also mixed ones – Great tithes and small tithes and the church, according to the

individual who enjoined [?] to church. They made a distinction between the rector, who does the work, and the vicar tithes. All kinds of tithes and arrangements were also known which do not interest us today. This system continued in Great Britain until less than a century ago. The matter came up next, of course, in Ireland. In the 90's the problem became very acute in Wales because the Welsh were not Episcopalians and they did not wish to support the English Church. We had the system here in Connecticut; everybody had to pay the church. But with the separation of state and church in this country, all that has disappeared. Now, gross produce is also a very rough criterion, because two pieces of land might have the same gross produce yet their expenses of cultivation may be different. They went over in the course of time, therefore, to the third system:

(3) This system takes as a criterion the fertility of the land or the site, good land as compared to bad land. The Romans had that system. Vectigal[186] was a tax on fertility, whereas ugatio was on quantity. Now, in this country, we find it also very common. South Carolina abolished the quantity tax and adopted this. A good many other states took it over. It was the first tax in Ohio. It is the system in China – the basis of all Chinese land taxation.
(4) This was based on the mode of cultivation – a little more refined system. You find this in Connecticut. Classification of land: mallow [sic: fallow?] land, arable land.
(5) The best system adopted was to take the criterion of net produce – gross produce less the expenses of cultivation – that is the general system. The English, French, Italians, and Germans follow this.
(6) On the other hand, a few countries like our own have a land tax on the selling value, and we find different systems also in Switzerland and in Canada.

Now, what is the difference between the 5th and 6th systems? One is a tax on net produce and the other is a tax on selling value. Which is the best? What are the prospects as a matter of theory?

Fundamentally there is no difference between capital and income. There is absolutely no difference, whether it is capital or capitalized income. Capital is a fund of wealth, income is the flow of wealth, so that you would say that there is really no difference. Practically, however, capitalization does not take place immediately in the case of land, as it does on the stock exchange. Why? Of course, the slightest change in the anticipated yield of a security will reflect in its capital value. In the case of land, this does not follow so conclusively, because you never are quite sure what the yield is going to be, how it is going to be affected by nature. It is a much slower process and in the interval there may be a great difference between a tax on yield and a tax on property value. The reason why we adopted the property tax is because the feudal system was entirely gone. Under the feudal system you cannot buy or sell property, all you can do is *usus* [*usus fructus*], and therefore land is worth only what it brings. Whereas, land is bought and sold here, and therefore it is much easier to measure it in terms of property value than in terms of yield. Wherever you find, therefore, the disappearance of the feudal system, there you have the beginnings of economic democracy, as you find in this country and Canada, and, moreover, you find a tax on property values, of land, the tax on income and yield values. Only in Australia have they adopted the British system.

Now, there has been a great movement, a double movement, which is very interesting. In recent years, some countries have a yield or income system instead of a property system. Property is a legal conception but not an economic conception. When we discuss property, we mean the selling value of land, the capital value of land. In England, therefore, we have a capital levy. Using the terms as you find then, you have to contrast a produce or yield tax with a property tax. In some countries movement is one way and in other countries it is the other way. Why do we find this? What took place in Germany? When I spent a few years in Germany I came across a situation of a house owned by a potato king. As land was taxed according to yield, he paid taxes on this land according to potato yield. But, in the years following, the town of Berlin reached to these places and houses were built and the potato king continued to pay taxes on the potato rate. You can, therefore, understand why this change is advocated. The same thing took place in Australia and New Zealand. There it took place when an attempt was made to adopt the single tax doctrine. In taxing land you must tax the land value. On the other hand, you find a situation in the contrary direction in Canada. In Canada during the last few years, especially after the war, the rents fell off very considerably, but because capital values are assessed values the whole revenue of many of these towns was derived from the tax on the land. They abolished the tax on the houses, the assessors were unwilling to reduce, and rents fell off largely. The owners of land were paying much more than the rent and, therefore, there was a decided movement in Canada, for this reason, to change from a property selling-value basis to the progressive value basis. In this country there has not been any such movement. Although farmers in the Northwest have had a hard time, they still have had to pay local property taxes. Their yields fell off and there came a great movement to bring about the change.

Criteria of Liability: A double movement is going on in this respect: movement away from a yield tax toward a property tax and movement away from the property tax to a yield tax. This is especially noticed in Canada. In this country, there has not been much discussion because of an inveterate tendency to leave the general property tax as the ideal of taxation. This has been the case in this country during the last century. But when the yield stops for a few years and market value falls, then we have a difficulty. Now, so far as probability to change is concerned, there is not much chance that a great change in the future may come so far as agricultural and city land is concerned. Especially in a prosperous community the lag between income and capitalization is a rather short one and on the whole the advantages of the property tax over the yield tax seems to be greater in rapidly growing countries in so-called unearned increments. Where land values are rising and land is held for speculative purposes, the unimproved property tax is better than a yield tax. In one of the Canadian towns a lot right in the town was kept entirely vacant for speculative purposes. Now, of course, a yield tax didn't touch it, a property tax would increase the revenue with every increase in annual assessment. Where you have conditions of this kind you still have this. After a while, especially when we come to the conditions in the old countries, the yield tax seems to be better than the property tax because it fits a little more closely. The problem with us now will not be an acute question. Even in those countries where they have a yield tax, the difficulty is solved by putting an additional tax on unimproved values. Our conclusion, therefore, would be that while, from the point of view of theory, yield is a great deal better, under the conditions we find in this country today probably on the whole capital value is better than yield and will remain so for some years.

Now, we come to the third problem of the land tax: What is the character of a land tax? What kind of tax ought it to be? We have a double point of view: It ought to a tax *in rem* or *in personam* [*in casoma* in original]. A land tax almost always begins with a tax *in rem* – a tax on the thing. And, only after the land tax develops into a general property tax, does it become a personal tax upon the land. Now, in this country, originally the tax was largely *in rem*, but for over a century now that tax has been mainly *in personam* [*in casoma*] on land as well as other property. In a great many cases allowance is made for indebtedness, money that a man borrows. Where the general property tax breaks down, and it has broken down in this state, the tendency has been for the tax to revert to the original form of *in rem*. It is found in this city and in a few other states. You don't find the name of the owner, you find the land. If the tax is not paid, the property is sold. Where you have a tax on capital value, the land starts out a tax *in rem* and ends with a tax *in rem*. That is simply a fact and the only problem about it is whether or not we should treat a tax irrespective of the indebtedness of the individual. The other side of the question, also interesting, is the historical development. Should the tax be apportioned or should it be a percentage tax?

In this country almost everywhere it is an apportioned tax, whether it is on land alone or on land as a part of general property. As regards real estate, we have this system in this state. Other states all have practically a general property tax. The amount of money that has to be raised is fixed in the budget. {Take the assessed valuation, divide one by the other, then you get a certain amount is ascribable to the other valuation, whatever it is you finally get down ultimately to the individual. The tax rate is fixed by dividing the amount to be raised into the amount assessed.} [The foregoing is utterly confused: The tax, or millage, rate is determined by dividing the amount to be collected, set in the budget, by the total assessed valuation. The resulting percentage, treated as so many mills per dollar of assessed valuation, is then applied to individual assessed valuations.] Taxes therefore are apportioned. One never knows from year to year how much he is going to pay. It is a political question. In a percentage tax, you start out with so much percent. In an apportioned tax, you know what [revenue] you are going to get; in a percentage tax you know the rate, you do not know what [revenue] you are going to get. Everywhere else in the world the land tax has become a percentage tax. In France we know exactly when that happened and even the so-called tax *réelle* [*reél* in original] has become a percentage tax, and also through all the other countries of the continent. Now the advantage of an apportioned tax is that it can be used as one of the means of securing the elasticity in taxation you need. There is a fight in Albany about this now. You need millions more or millions less; whereas a percentage tax, it is true, lays a low rate, you cannot be sure what an increase would bring. With the federal income tax, for instance, change the rate and sometimes there are discrepancies amounting to several hundred millions of dollars. The advantage of a percentage tax is that every taxpayer knows beforehand what he is going to pay. On the whole, the tendency is in the direction of percentage taxes.

Now, then, we come to the fourth problem connected with the land tax, the important question of assessments. How is it assessed? There are quite interesting miner problems, but the main problem would be the method of assessment. What is this method? Of course, in this country we make out a tax list or roll of property. That is assessed in some states; we have double inventory in some other states. In Vermont this is called the Grand List. You go down to the City Hall and find exactly out at what value certain land is assessed. The important question is, how are you going to reach the assessment? Whole valuation? Of course, there are different ways of doing this. In the case of a tax on yield or produce, you must have cadastre. Every piece of land is put down after careful inspection, and how much it is going to yield; it takes a long time to make these surveys. It is a very difficult thing, and by the time you are through with it you must begin all over again. In the case of a property tax it is much simpler, as you have valuation from year to year; property is bought and sold in this country. Even in the farming district it is not so very difficult to form some sort of idea as to what property is worth. It is far more difficult to form an exact idea as to what a piece of property is worth. When the Woolworth building was built on that piece of land, it is impossible to say how much more the land is now worth. The mere fact that it increases the value of land, that requires very careful and technical appraisal. In this country today things are done in a very haphazard way; they only guess. Those methods have been introduced in our cities but they have gradually been improved. New York City is at the head of the cities of the world for assessment rates. We have very few examples of that kind. lt is largely due to the work of one man about a generation ago, Mr. Lowson Kirby.[187] He was the head of the department of assessments, a very intelligent and active man, and his activity resulted in an immense step forward. There is only one other place that can compare with New York, and that is the city of Cleveland, where they also had a very intelligent assessor. The community would not let him go. This man is Mr. Zangerle [Zangarer in original] of Cleveland.[188] What Kirby did gradually had its influence in other places. He introduced block and lot maps. Every piece of land in the City of New York and the boroughs was mapped. In the City of New York there was forced upon us the system of the 25′ × 100′ lot. According to this plan it was assumed that traffic would go from West to East. Of course, things turned out the other way, which was the main cause of our congestion. That idea of having a lot of 25′ × 100′ was responsible for all of our evils. How can you build a house on 25′ × 100′ and have enough light, or other conveniences? Other cities are much more fortunate, Baltimore, for instance, etc. In Paris there are no definite lots; land there is sold by square meters. In New York, how are you going to estimate the exact value of a lot? Some lots, on account of the formation of streets, etc., are more than 100′ deep; some of them are 75′ or 80′ deep, how are you going to estimate that? And then you have an inside lot, the inside tenement houses. What is the value of an inside house compared with an outside corner lot, inside lot and outside lot? So there have arisen all sorts of technical rules: 1,2,3,4 rule. The first was considered 40% of the entire value of the lot; the next, 25%; etc. Then another gentleman developed another rule, percentage or difference. One of our older students who went into active political life, Mr. Neal, also worked with Mr. Hoffman, called the Hoffman-Neal group. The first lot, 44%; second, 60% etc., that is how they worked it out. Our New York rule differs from the Cleveland rule, which gives more value to the first 50′, and there are also other rules.

The difficulty in getting exact assessment on a piece of land in a city sometimes so baffled the assessor that private companies entered the field. In this country books have been published trying to provide more exact rules for ascertaining assessments. At other places, things get more complicated; where you have houses, the great factor depends on the kind of buildings for the value. There is room for necessary and painstaking accuracy. Especially in the South and West, accurate assessment is still in a very primitive stage, in ordinary conditions almost as bad as in our county districts, where they guess the valuation. This is a real science in itself, and it has been developed more in this country than anywhere else. Japan sent men over here to study our system. The method of assessment is a question of great importance.

The second question under this head of assessment is frequency. How often ought property be assessed? The ordinary term has become annual. We have a two-year term. There is a three-year term in Washington, D.C.; four-year terms in Delaware, Mississippi, and North Carolina; a five-year term in Virginia; ten-year terms in West Virginia and Ohio, and also we have a twenty-year term. In this respect, we get almost to the case of the potato king. If you have property assessment you ought to have as frequent assessments as possible. They ought to have certain rules; for example, when skyscrapers were built around my house, I protested that this prevented the air and spoiled the scenery, and allowance were made and the house was reassessed. Assessment is also rather important in connection with the budget, such that the period of assessment ought to be a year. The last question under this head is perhaps most important, namely, the rate of assessment.

You assess your property and it is worth so much, but, how much is it worth for tax purposes? Ordinarily in most of our states tax officials and collectors levy the tax on the full value. For a peculiar reason the taxable value has continually been diminished: that is largely due to several states having the apportioned rather than percentage form of taxation. If New York tries to raise 20 million dollars by direct taxes, you divide it according to the assessed valuation in the counties. Now, the smaller is the assessed valuation in each county, the less taxes are put on that county, hence the race to reduce the assessment, because of the fact that when one county assessed at the full rate, but the adjoining county of the same size assessed only at 50% of the value, the latter county would pay only half as much tax. Therefore, there has been this race to reduce assessments. In different counties they have different percentages, such as 9%, 8%, 7% etc. Some counties have 20% or 10%. In order to get rid of this problem, they make all have a uniform 20% or 10%. For instance, in Illinois the rate is 20%. This fact is surprising to a foreigner who does not know the exact situation. But this is actually different, as the property is assessed at its 1/5th value. In New Hampshire, they assess at a rate of 50% of every $100.

This has brought about two great evils. It is responsible for inequality: a man with the same property or income in one county may pay only half as much as a similar farmer in another county. What is true here is true elsewhere. When you have gotten away from equalization you try to come back to it. When, therefore, the income tax was introduced into this state, I suggested that the income tax should be divided among the localities. I said that instead of dividing the income tax as you divide the property tax it should divide the benefits according to assessed value; every county will receive according to the assessed value of its real estate. That is one of the ways by which we stopped that practice. The other was still more efficacious. About forty years ago, I suggested giving up the raising of state revenue through the real estate tax, limiting the real estate tax to local governments: there would not be a race for reduction.

One of the great reasons why this movement came about is the question of the separation of state from local revenues, which is an active political question today. Keep the real estate tax for local purposes, so they don't need under-assessment, and add to that the idea of distributing revenue from the income tax according to the opposite principle, and you get rid of the difficulty. Only a few states have seen this difficulty.

The only other question under this head is whether your state has proportional or graduated taxes. You find graduated land taxes in Australia and Switzerland, and you find it here on the statute books in Oklahoma. From a fiscal point of view, a graduated tax is absurd if your tax is *in rem*. Progressive taxation is the last word in the theory of ability pertaining to the individual. When you apply a graduated tax on land you get all sorts of absurd results. For instance, a poor man has a piece of land worth $1,000, all he has in the world. On the other hand, a rich man has land worth $500, but the rich man has millions. The poor man will be taxed much more than the rich man; there is no correlation between these two cases. The only reason why they have it, as in Australia, is because in those countries, where the tax is a personal tax, they attempt to achieve non-fiscal results, social results. Australia did not want to have an aristocracy. That of course, may be a perfectly desirable thing, but from a fiscal point of view the graduated tax is really absurdity.

We come next to the question of equalization, which with us is a very great problem; in France they call it "*paritation*", where they attempt equalization. We have a perfect chaos of efforts in the system in this country. In some states we have a State Board of Equalization promoting equalization between the counties, villages, and local units. Some states apply equalization to real estate, some apply it to personal property, some apply it to all, and some to aggregates. So you can easily see what a multiplicity of methods you have in case you want to find out what the questions are. The trouble with those state or local boards is that even though they have the legal power, they haven't either the knowledge or, what is sometimes more important, the courage to strike at the root of the difficulty. The tendency to undervalue under our general system in this country is irresistible, because you are putting a premium on the valuation by making the legislative tax a state tax, and for that reason so many states find this under-assessment inevitable and make under-valuation legal. The remedy is not through these boards, but through some much more radical methods.

The next question is that of classification of the land tax. The reason for that is largely of a practical character, that if you attempt to tax all land according to our system in this country, taking the selling values, then you get some very bad and dangerous results in certain kinds of land, lands which are not used on account of being very poor for agricultural purposes. In some states, however, you find a classification which is much more detailed and which is largely introduced for administrative reasons. Take, as an example, the state of Idaho. The law there requires assessors to distinguish between city and urban land, between arid and fertile land, between ordinary agricultural and irrigated land, between line-built [?] land, dry-farming land, meadow and pasture, between desert and waste land, between mineral and timber land, and between various kinds of timber land, cut and burnt timber land. Even city lots must be distinguished between business and residential sections, and various other kinds. These are designed rather to aid the assessors in their task to determine the adequate values. You find examples of this kind in a number of states, although nowhere else quite so detailed. By far the most important thing is classification where the object is to get away from the common method of taxing land according to selling value.

The situation is different when we consider other kinds of lands. There are three classes of land where there have been great change: (1) Forest lands, (2) Mineral lands, and (3) Waterpower lands. Why do we find new legislation about forest lands? For the simple reason that if you tax land according to its selling value, and if you have a growing forest on that particular land, it will increase in value from year to year, reaching its maximum value when the forest matures. In this case the chief value of the land is the wood on the land. Now, under the system of general property taxation where assessment is increased from year to year, the taxes grow without any corresponding change. However, the owner has got to wait until he markets the timber. It takes anywhere from 50 to 75 or more years for timber to grow. In other words, the pressure of the tax is almost irresistibly for him to cut the trees before they mature, as for instance, when they are only 20 or 30 years old, and sell it to the wood companies. Therefore, we can clearly see that one of the main causes of deforestation of our land is our system of taxation. When we were developing our general property tax we did not think of this problem and the result is that we are slowly moving to the adoption of the European method. In Europe the system is largely that of a yield tax: tax the land when you sell your timber and don't tax it in the interval. The trouble with that system in this country is on account of our peculiar arrangements in the relation between local and state governments: so much depends upon the annual revenue of each locality. You have to keep up with your schools, keep up with the roads to a certain extent, and for reasons that we will see, it is the real estate tax that will be serviceable and the personal property tax unserviceable. If you are going to wait until the timber is marketable, and if a large part of land is planted with timber, this will not be practicable, and, therefore, we have adopted a kind of compromise system. This is especially true with our recent legislation in New York State and the state of Massachusetts. In a few states since about 10 or 15 years ago, we have attempted to combine the yield principle with the property principle and introduced the Massachusetts commutation principles. The idea of those laws, or of similar ones in different states, is to abandon the property idea but keep some of our tax on the {par value of the land irrespective of value.} In Pennsylvania and Massachusetts, this has not been a great success, the object being to {reverse it to lands}. The whole problem of taxation of forest lands is discussed by Professor Fairchild, who had the reputation of bringing the matter in front. Endres is another man who has a book of from 600 to 700 pages dealing with taxation of land all over Europe. Fairchild utilized the results of Endres and is responsible for popularizing the influence of European scientists in that respect. We now have to wait until the committee reports appear. It is going to be a kind of combination between European and American systems.

When we consider the mines, the problem here is very acute, because so many states consist of lands having mines. Now, the reason why the older system proved be inadequate as the general property tax was because of a little different situation. The trouble was rather the other way around. In the case of forests, taxes were so high as to cause deforestation; but in the case of mines, the ordinary local system of assessment left the mine virtually untaxed, so that it did not bear the proper share of the burden. Because, in the first place, they did not know the amount of mine underneath the ground. As a consequence you have here a situation where, in many states, the total assessment of property value was less than it should have been [it was taken from year to year, in original]. This was due to inadequacy, and a three-dollar-a-day man was too ignorant to know what these mines were worth. The result was that there was no equality, and equality was not brought about between mineral and agricultural land. They then resorted to a system of yield tax, and, as a result, there was a very interesting development. At first, they took the gross yield. Of course, this was a very rough method and they made certain deductions to get the net output. In some states you have a combination of gross and net profits {and then again in some states, Colorado, for instance, and others}. (You have two or three good books on this subject.)

More recently, there has been an interesting development in the other direction. You find that in New Mexico, Michigan and Minnesota. When Professor Haig [Hague in original] was called to New Mexico, after careful study he made one of his best reports and he came to the conclusion that on the whole the difficulty could be met with by favoring the system of net earnings. You see, it is the same kind of difficulty that the income tax officials in Washington have met with in trying to get an income tax for mines. And, Professor Haig [Hague in original] concluded very wisely that if the state would be willing to employ high-salaried experts, mining engineers and expert geologists, that it would be easier for them to come to an approximate conclusion as to what the selling value of the property will be, rather than to conclude what the income was. The entire question is to ascertain as to how you can practically achieve equality. In some respects it is easier to fix upon property values, and in other respects it is easier to fix upon the earning value. Sometimes one is true and sometimes the other is true. In the case of mines, the experience has been that they really get fairly good results by utilizing a general property tax, but in a way that is very different from what they have done in the farming district. Very detailed and thorough scientific investigation is necessary to verify this, but, of course, even then you are not sure because nobody can tell exactly what the value of the property is. There are all sorts of difficulties, but the experience in some of our states has been that you can get on just as well by expert assessment of selling value as by expert assessment of yield value, if we apply to yield and earnings of the property the same degree of skill that the federal government applies to income. It really is a choice between an inefficient method of assessing property values and a similarly inefficient method of reaching earnings. If you are going to have inefficiency any way, it is better to take the earnings. But in most of our states they would rather have an [in]efficient value of property than an inefficient value of yield. But if you have [in]efficiency of property value, earnings will be better, so that it all depends on what part of the country you come from. This problem, no doubt, will be a very acute one in a few years.

The same thing is true in Texas [Taxes in original]. We know that pretty well in the case of oil when it is exhausted, better than in the case of mines. Finally, the latest vestiges of the problem in this country have come from the Western States with reference to water rights. How are you going to measure the worth to the owners of the water rights? and the value of the land? Property valuation does not come near doing the job because the earnings of a crop are not capitalized into the value of land, they are capitalized into excess of crop. A great many other things come into this question. Therefore, in the case of water rights, the whole tendency to utilize earnings and income is very much stronger. The net result of this whole discussion of classification is that we are getting away from the system of selling value which is still utilizable for city lands.

Now we come to the next question, which has also entered into politics: The question of the exemption of mortgaged debt. What shall we do with a piece of land that is burdened by debt? And, how are you going to get equality between a man who owns a farm free of debt and another man who owns a farm by debt? That brings about a very great problem of equality and we have now three, four or five systems used in this country to deal with it. In some states they exempt mortgages completely from taxation. If a 10,000 dollar farm is mortgaged for 5,000 dollars, then his tax is only on 5,000 dollars, but that implies a reduction on the real estate tax. But the land tax is a tax *in rem* and the personal situation has nothing to do with it, therefore you cannot apply this deduction for debt. The arguments going on over the tax *in rem* are very strong. That solution would be possible only where the tax is a personal tax, and we don't find that in this state. Some states don't make any allowance for debt at all. They run up against the objection that so far as the owner is concerned it [debt] is inevitable. We allow in this state an exemption on personal property. Other states pursue a different method. They lay the tax upon the mortgage as an interest in the property at the place where the property is located. For instance, in county A, there is land worth $10,000 which has a $5,000 mortgage on it. This $5,000 mortgage is owned in county B. Now, instead of taxing the man in county B, they tax the man [the mortgage holder] in county A. He has to pay all the tax in A and deduct part of the payment in the other county. The trouble with that is that if that man knows about it he is going to try to make up for that. Therefore, some states add to this a provision that the [mortgage] owner is prohibited from making any contract with the borrower to make him pay any more. You can very easily see and understand what this amounts to. It took a long time for the farmer to learn this; he has learned the reason, and that is why you don't tax the mortgage at all.

Some states permit only a certain amount. For instance, you can only exempt $700, and in some states this applies to a certain percentage. Other states, which still need some money and don't want to exempt the mortgage entirely, reduce the tax [rent in original] on mortgages, such as in Pennsylvania. This comes under the heading of personal property. Some states, like New York, have introduced a mortgage recording tax with recording at certain amount. We had that at one time but abandoned it. So you see that is a very practical and often burning question in our states. But the simplest system is to exempt mortgages entirely. When you are dealing with tax question the number of practical questions is very great. Every one mentioned is in politics and remains in politics.

Another question is the exemption of improvements on land from land values, which is one of the most important issues. The problem is multiple and there are several questions: (1) Should improvements be exempted in general? Should there be any tax on a house? (2) Should there be any special land value taxes in addition to the general tax? Should you have, over and above, a tax on land values? (3) Should land be subject to a so-called unearned increment tax? In some parts of Canada, also in Australia and in South America, you have either an exemption in total or in part for houses, or you have a special tax on land values. In other parts of the world, such as in England and Germany, you have a system which was developed in China.

I want to discuss now the principle underlying exemption. The reason why the demand sometimes is made that buildings should be exempt from taxation is either because the single tax movement has made some progress – that, of course, is of little consequence – or because of the application of the theory of incidence. You remember from the theory of the incidence of taxation that a tax on land cannot be easily shifted, whereas a tax on a house can be shifted. Therefore, the argument is: If you exempt houses from taxation, rents will be cheaper, living will be cheaper, there will be less poverty and it altogether would be a fine thing. But, as regards that argument, you must remember two things, that if the government needs a certain amount of money and if you exempt houses, it simply means that the tax rate on the land will have to be somewhat higher; and those who believe that is a good argument in favor of conditions existing in this country today, must consider that you would have as a result what you practically find in Canada. This was tried in Canada and did not work. Municipalities went bankrupt and had to impose other taxes. In the second place, people who dwell upon that argument forget their general economics, which is that there is a certain relation between wages and salaries and the cost of living and that in the cities where rents are high the wages are high. It is simply an adjustment of the whole thing to the general economic equilibrium. We had that problem in 1915, when we lowered the tax rate on the houses of the city. It was on this that Professor Haig [Hague in original] did his admirable work. He draws the conclusion that in New York it really created more harm than good. In a small city this will be possible, but in a big city it would be impossible. The conclusion would be that in our modern American cities at least, there is no special reason why improvements should be exempt from taxation.

We spoke about the taxing of houses and also of the possible desirability of a special additional tax over and above the real estate tax, either on land values as such or on the unearned increment, that is, on the increase of land value from period to period. We have examples of both kinds. The unearned increment tax gets its name from John Stuart Mill, who first called attention to the growing value of land increment. It was his influence, you remember, which was largely responsible for the adoption by Australia and New Zealand of the policy of leasing rather than selling land to take advantage of this unearned increment. From that point of view we found that, as a result of experience in the City of New York, that argument is not very strong. Our experience in this city was that the whole community gains more by selling the land and securing an increment to the taxes on the greater value of land.

However, you come now to the other question, whether there should be a general tax on the unearned increment of property in the hands of private individuals. The first illustration is found in China in 1898. It was due to neither Henry George nor John Stuart Mill. It was decided that it would be wise to leave certain a portion of the increased value of land to the government. So they put that in operation in 1898 to help build the modern city of Kai Chau[189] [Kia Chau in original]. Later on, not only did colonies adapt [adopt?] this method but also a Henry George Union was formed. The tax swept like wild fire in Germany. These taxes went to 20% or 30% of the increased value, {if the increased value amounted to this}.

After a while, however, the idea gradually gained ground and the question was asked, why limit this unearned increment tax to the towns and localities alone? The argument was that the owner of land enjoyed this unearned increment, this increase of value, which was due to the activities of the population about him in the town; therefore, the money ought to go to the town. It was pointed out that this did not depend only upon local but also state operations. As long as this was unearned, why not make it wider and make it a state tax? They did not stop there, however. In New York City the increase was due to the tariff and all sorts of laws that are made in Washington.

If this is so in Germany there ought to be an Imperial tax instead of a local tax. Of course, they carried it one step further to the international questions.

They took still another step by this time. As the interest in single taxes became so great, the public began to ask, is land the only thing which earned unearned increment? and, are there not other things which received an unearned increment? You buy stock, the prices increase, and this increase comes from the increase of population. The conclusion is that if the unearned increment is not limited to land, why should the tax not be generalized not simply from the point of view of geographical conditions but from the point of view of economic conditions? Sure enough, they levied in Germany an unearned increment tax on everything. During the war sometimes they took up to 100% of the unearned increment. You see they got away from the land tax, but when the war was over you had a general unearned increment tax. By that you came very close to a great problem. In Germany the great question then arose, is it better to have a tax on the capital income, as we levy it in this country, or is it better to restrict the income tax only to non-capital income, as in England? What they did in Germany was very interesting. They decided to abandon the whole idea of unearned increment, therefore, they abolished the Imperial tax.

This unearned increment tax argument started two decades ago as a small tax on yield, which was a negligible amount. So there is the history. They tried this in England in 1909 in Lloyd George's budget, which led to a revolution and a change in the constitution. {The land taxes of England were four fold, it included good many things.} The chief thing here was a tax on the unearned increment of land and of the land value. I want to call to your attention that, of course, this started a resistance by the land owners, and the administration did not find any public sympathy, as it costs more to make valuations of the land than it was worth. Then the whole system was abolished, and England today has no land tax.

In this country, various commissions have dealt with this problem. In the City of New York or in Chicago, there is a growing feeling that {is really in favor to make free of all these growing advanced values}. We all know that fortunes were made. One of our professors wanted to make some money. He borrowed $100,000 and invested it in corner lots. Soon he sold them for $400,000, and now lives in comfortable circumstances. Of course, anyone with a little foresight, especially if he gets in this, can do the same; and, as a result of such speculation, millions and millions of dollars have been earned. Land values now amount to over 15 billion dollars, and some people say, let us utilize the system of special assessments. That is very much discussed about subways. Still, you can only go so far as putting a small tax on the unearned increment. A special tax on value, even an increment tax, is not needed as much in this country as it is in foreign countries. Our general property tax takes account, in part at least, of the increasing land values; but abroad, where you have the capital tax and yield tax, the situation is different. But, at best, in this country from the point of view of the unearned increment tax, it is not likely ever to attain the proportion that some of the enthusiastic advocates have advanced.

Now we come to the next point, which is still more important, the ninth question, as to whether the real estate tax ought to be a state tax or a local tax. Of course, they all started out with a local tax and then it became a state tax. Historically it is interesting that Germany started the relegation of the land tax to local purposes and one or two countries followed suit. England followed this. England has a wonderful feeling for what is politically fit. As a matter of fact, England, for over a century or century and a half, has limited the real estate tax to a local tax. In this country, the general property tax is the main source of both state and local revenues; the chief consideration of the real estate tax is its utilization for state as well as local purposes. About a generation ago, I tried to bring about what is known as the separation of state and local revenues, in which the real estate tax is used only for local purposes. We continued this for several years. California has had this for a long time. But certain difficulties in our arrangement, when times of stress came, brought about the reintroduction of the whole system. You remember that government set to look forward to a complete disappearance of real estate from state taxation. But even that very slight tax maintains the abuses and difficulties of making under-assessment of property your last reliance. The conclusion is that the wise policy is to do what other states have done, that is, to utilize the real estate tax only for local purposes. There is no doubt that theory and practice tend to conclude that the real estate tax ought to be local, as it is in England and in Germany, as it is in California and as it is getting to be in this state.

The last question in connection with the real estate tax concerns the rate of the tax, whether there should be any limitation of the rate. Where you have, as in this country, such exclusive reliance put upon the property tax, the more money you need the higher the rates will grow. The public becomes frightened and starts passing laws limiting the rate of the tax, first for state purposes and then for local purposes. They even go so far as to put this in the constitution, because they don't trust the legislature. For instance, the state of Kansas had a limit of one-half of one percent; in other cases, three or four percent. The most outstanding example of the difficulties of this situation is found in Ohio. Any one of you coming from Ohio knows that the question turned around the tax limit law known as Smith's law.

Now, what are the objections to the limitation by law of the tax rate? In the first place, it is like having a rubber air-filled cushion which, when pulled down at one end, the air will go to the other end. What I mean is this: If you need more money and limit the tax rate, you are going to borrow and the only result will be the increase in the city's debt. When the legislature woke up to that fact they said they would {leave everything}. That is all right when expenditures are extravagant, but if your system is not so bad, what is the difference? On account of a lack of funds in some states, schools were closed, and for eight weeks streets were not swept. If, they said, you can do that, you might as well go back and give up the government. If the expenditures are desirable, the way out is not by limiting the tax rate. One of the students took up the question of the limitation of the tax rate in Ohio and came to the conclusion that you can not solve anything by simply passing a law; the way out of the difficulty is very different. You see, at once, that there are many very important points connected with the real estate tax. You might ask, how much limitation should be done? In the various cities, for that matter also in the country districts of the United States, it is only in the last two or three years that agricultural institutions are taking up this problem with doubled energy, probably all farmers. We have no time to go further into this question.

## 68. **House Tax**

This is the other side of the real estate tax that is important. Houses may be taxed either on the owner, as we do, or on the occupier. If it is taxed on the owner, you can tax it upon the value of the house, as we do, or upon the rent which comes in to the owner. If you tax upon the occupier, you can tax it either upon the rental value of the house, as they do in England, or you can attempt to take the rental value, then the expenditure made by the occupier according to his ability to pay. Instead of taxing income, they take the rent paid for an apartment and multiply {it with so much}. Now, limiting ourselves to our own problem, we have to go into the question of the occupier. How far does the tax affect the occupier? We will not go into these problems, as they do not exist with us. The matter with us is that the house is always made a part of the real estate tax, and we don't distinguish the house from the land. Therefore, everything that applies to land equally applies to the house in this country. We have had, however, an additional kind of tax upon houses. The federal government, you remember, had taxed houses and the occupier back in the 18th century (1798), when we needed to build our navy. A great investigation was made by Mr. Walker,[190] Secretary of the Treasury and two or three plans were presented. Finally, in 1798, they adopted, for the first time in this country, a progressive tax – a graduated tax. They were opposed to the French politically but they took their economic views. Houses were taxed according to their value, two mils for $1000; the highest point was reached at $30,000; at that time they did not have any houses worth more than this, which was [at] 10 mils.

We have had, at various intervals, suggestions to have a separate tax on houses from the point of view of expenditure rather than property. It was suggested to introduce a tax such as was adopted in Canada but not in this country. A few years ago we had our great commission, of which I was the chairman. We limited the state income tax, what we called an occupation [occupier] tax, a tax that the renter paid. But, of course, when you already have an income tax, this is far better; any other movement has never succeeded in this country, because it is not necessary. Expenditure is not a good criterion for ability to pay; it is not likely, therefore, that we shall ever have a tax on houses apart from our real estate tax.

## 69. II. **Tax on Personal Property**

This, of course, has given us a great deal more trouble than the tax on houses. What are some of the reasons? You can tax personal property in four or five different ways:

(1) Like the general property tax, which is as a real estate tax.
(2) You can levy a tax on the separate classes or kinds of property, such as intangibles.
(3) You can have certain forms of personal property, as is typical of an expenditure tax, luxuries and such, as we did during the war, such as taxing automobiles, etc.
(4) You can tax the yield or produce, or personal property as part of the income tax, as we do in this country.
(5) Of course, you can tax things that you own – a piece of property, an indirect tax or taxes on commodities. This last subject we shall deal with in the next lecture or two. Today, I will limit myself to the second and the third.

## 70. **Tax on Tangibles**

Personal property, from an economic point of view, is capital and consumable goods. But virtually, to all intents and purposes, we mean by personal property in this country today that kind of capital value which is to be distinguished from real estate. That takes in furniture, pictures, the individual tools of workmen, the machinery of manufacturers, the business of corporations and the securities and the cash owned separately, etc. In other words, it includes all productive wealth but it includes also consumption wealth – it includes watches, automobiles, etc. That kind of classification does not help us very much. For all practical purposes, the classification used in this country is that of tangible and intangible personal property. Tangible are things that you can touch. For a merchant, it is his stock in trade. For the farmer, it is his tools and implements, chickens, cattle, etc. For the city dweller, it is his furniture. For the individual, it is his jewelry, etc. On the other hand, you have the intangible things, which are composed chiefly of mortgages, corporate securities, stocks and bonds. You mean virtually the securities. Now, when both of these are taken together in this form, personal property slips out of the assessment tax. It is exceedingly hard, even in this case, to determine the value of personal property. For instance, in Philadelphia a census of watches was taken, and among a population of over a million only 30,000 watches were found. It is exceedingly hard because you cannot enter into a man's house and you must guess. The only person who suffers from attempts of this kind is the farmer, because you can see his tools and count his chickens. Even in the intangibles you find the farmer suffers most.

We see, therefore, that there is considerable difficulty in the assessment of tangible property. For one reason or another, the tax on tangibles has been abolished in this state. Our total income from intangibles has been, in this state, several hundred millions, but on tangibles only about two millions of dollars. In some counties and states, we have no returns at all.

Also, the chief reason why the case of the tangible tax is not theoretically important is that all these tangibles are really not a source of revenue but a source of expenditure. You get a psychic return from your jewelry but do not get any other thing. So that, even theoretically, there is very little foundation for it.

## 71. **Tax on Intangibles**

This consists chiefly of mortgages and corporate securities, which utterly elude the assessors. It has been of very little importance from the beginning because of the great inequality in some states. Office boys used to do this; they were the assessors of 100,000,000 dollar tax. Another way was to swear off. {But if a man is assessed a large amount, but if he should own 10% and if he is honest man, particularly honest, he can not swear off.} I myself was caught in this way. These cases happen all the time in states where you have this system. The tax rate in the city amounts to 3%, income from bonds may be 4%, and if you are caught you must pay 50% of this income. This has become such that there have been three different ways that the American states have gotten rid of the problem. In Europe it disappeared two or three centuries ago. They do not have this in Europe. In this country, now, there are three methods:

The first method is to try to induce the owner of a security to pay by reducing the rate. Instead of taxing 3%, tax 3 or 4 per mil; that is known as a low rate on intangibles. A number of our states tried this and in almost all of them it was found impossible. Where the constitution did not prohibit it [exist in original], as in New York, or where it was changed, then this avenue only was open. The first year they did not get any more revenue by a 3 mil tax, still a great deal more was returned {for assessment}. It only lasted for a short time. After a few years, the whole story repeated itself and they said, why should we pay even this 3 mil tax? The temptation is too strong when the custom is no longer to obey the law. While we still have a different system and the situation is bad, yet most of the states {have gotten over the second or third plan}.

The second plan is to abolish the tax on personalty and to put in its place a tax on business, because most of these securities are owned by businessmen or professional men. That was advocated very much in this country, but adopted in Canada mostly in Quebec, etc.

The third plan, followed by Massachusetts and New York, and by some other states, is to abandon the whole tax on intangibles and to introduce an income tax which separates income from personalty. Massachusetts had a great fight about this. However, later on it also went to an income tax. Professor Bullock was a great advocate of this. Where states for economic reasons are not yet ready for the personal income tax, although they may be later for a business income tax, then the low rate of an intangibles tax is to be recommended. Even then, it does not put the city man even with the farmer, as the farmer still has to bear most of the tax, because his taxes are on tangible personalty. But, in one way or another, these are beginning in this country, as anywhere else, with the exception that in some states it is required to register mortgages. Even this has disappeared. Let us now take up the tax which is taking the place of the personal property tax, the business tax, although it has some independent history.

## 72. III. **Business Tax**

Of course, you can tax business through the property tax or tax the stock. You can also impose a stamp tax. Tax through the income tax, levy an excise on the produce of business, or you can tax the business as such, a tax not on the principal but on the business, a semi-personal tax. This is assimilated to a land tax. Now, then, from the point of view of a tax on business, there are four chief problems that ought to be discussed:

(1) What are the subjects of taxation? You tax the business, but what kind of business? You find in this country the business of individuals, or the business of firms, or partnerships and business of corporations.
(2) The question of inclusiveness of the tax: A certain kind of business, such as bank business, railroad business, or a tax on all businesses.
(3) What are the different criteria of the kinds of taxes imposed upon individual? Here we find four chief categories: (a) What you find in this country, a license or privilege tax. (b) You find what you call in this country, special taxes or privilege taxes. (c) We have these general business taxes, under different names. The French call it *patent*, the Germans something different. (d) You may have what we call in this country corporation taxes or the franchise tax.
(4) In this set of problems we find six criteria: (a) You may have a flat tax. (b) You may have the tax on capital. (c) You may have the tax on gross output. (d) You may have the tax on gross receipts. (e) You may have to use other indices. (f) You may tax the business on the net receipts of profits. You have great variety.

## 73. **License Tax**

We will say a few things about the license tax. It originated in England in the 17th century; certain trades had to take out a license to do business. These early licenses were on drinks and then on trade in general. For instance, the tippling license started in 1552. Later on came the other licenses in quite a number, such as developed in England as licenses for coaches, state coaches, pawnbrokers, shop keepers, auctioneers, etc. In the year 1888, all these licenses were turned over to local government {and all known local taxes on licenses}.

In this country we have had examples of federal, state and local licenses. In our federal legislation these have been found chiefly in time of war. In the war of 1812, in the civil war and recently. Sometimes they are called license taxes, sometimes special taxes. We call them today special taxes. But there are only a very few cases, such as dealers of oleomargarine, etc. In state and local taxes they have assumed a greater proportion. Before the passage of the 18th Amendment there was a tax on liquor, and many states derived their chief revenue from it. With the passage of the 18th Amendment we lost about two billions of dollars. We could do away with our entire income tax. This was a great sacrifice. In many countries these liquor licenses are still very important and are a chief form of revenue. The different special taxes that we had during the war also were in great number. In this state, however, we still have a whole mass of such taxes known by various names. Some of the Southern states call them occupation taxes, some call them privilege taxes, and so on.

Theoretically the difference between a license and an occupation tax is that in the case of the license tax you have to pay the tax before you do the business, whereas in a regular business you pay the tax after you have done the business. The greatest examples we have of a business tax are in France where, you remember, after the Revolution they abolished all business taxation. Business became one of the four sisters, another one was land. Giving patents in business became a chief source of revenue. It is a very complicated system composed partly of proportional taxes and non-proportional taxes with great detail. The system, however, was abolished during the war for both state and local purposes. You still have this in France. From France it spread to all of Europe – Germany, Italy, etc. You still have it in the other parts of the continent. This tax is very important. It is not on a man but on {the business and combination of rental values}, and they are classified accordingly. They have this system in Canada and also in Japan.

With us, however, business, as you know, takes the form primarily of corporate activity; therefore with us the chief illustration of a business tax is the corporation tax. Though not exclusively: as one of the suggestions in the Mayor's meeting of last evening, the committee accepted the recommendation of adding non-incorporated businesses. There is no doubt that there is a general movement to extend the corporate tax to non-incorporated institutions under the influence of the income tax: for federal purposes, the income tax is paid irrespective of the individual.

The corporation tax is a very big subject. You remember, I devoted to it three long chapters in the *Essays*. Those who really want to get the essence I shall refer to the following pages: pages 142 to 150; 166 to 170; 177 to 182; 218 to 220; 230 to 250 and the last page. All I am going to try to do now is to deal with it from the four points of view:

1. Historically: The earliest corporations in this country were banks, roads, canals, etc. We began with taxes on railroads, and it was not until the middle of the [19th] century that the corporation tax first came in Pennsylvania. {You still find the tax in some of our less developed states.} The earliest taxes were on railroads, public utilities, telegraph companies, then also other public utilities. With economic development, taxation was gradually extended to all other corporations: trading corporations in general, factory corporations, business corporations and finally the general corporation tax.

Now, as regards the facts: The facts, today, of different classes of corporations give you a true picture of the historical development of business taxation. Take the railroad, for example, which started in with the attempt to tax railroads as a general property. One county suspended the right of way of cow pastures. The matter of local assessment of railroads soon proved to be very defective. A few of our states, like Wisconsin and other states, fit railways into their systems of property taxation. {But state assessors in Alabama took the value of railroads and they utilized more a recondite, more settled tax.} So, it really amounts to the same thing. The great difficulty in this country is that where you tax a railroad as property the franchise problem arises and it is only in the United States that this problem exists. This has given us more difficulty than any other tax: {when you take railroad you cannot get tax, you have franchise}. (This matter we shall take up later; this is one of the most difficult of all the problems.)

We have, as a result of all the complexities, developed three kinds of franchises:

The first is the right to become corporate. A good many so-called franchise taxes are nothing but incorporation fees paid for the privilege of becoming a corporation.

The second kind of franchise is the franchise to do business. That kind of franchise is the most important kind. Every possible way that human ingenuity has devised has been tried [to place a value on it]. Whenever assessors try one way the big corporation lawyers say it is not the right way, though they do not show the right way. If you look at our state and local reports, you will find every possible theory that you can think of, as to how to measure [the value of a] franchise. A certain tentative decision was made by the Interstate Commerce Commission [interstate commerce in original] putting it at 9%. This would have caused the bankruptcy of all the public utilities. It depends how you are going to estimate depreciation in connection with a franchise. {All I can point out is that as soon as you abandon the property criteria and accept some other criteria such as gross receipts or net receipts, when you get rid of all these difficulties and only in this country the property idea that we have leads us to difficulties.} Therefore, in this country as a whole, the tendency is away from property and toward earnings of some kind as the value of franchise.

The third kind of franchise is the most interesting and difficult of all franchises, e.g. the special franchise. This came about as follows: In this country, and in this state, you can deduct debt from personal property, but you cannot deduct it from the property of a corporation. When local taxes became very high and when taxes were 3% on their personal property, they also found the tax very high. They then themselves issued large mortgage bonds and the issue amounted to more than the outstanding stock. Under the law, since they can deduct debt from personal property, they found that they have a minus property tax. In the government in Albany at that time, a Senator devised this plan. He said, we will not tax a company on its property, we will tax it upon its franchise, not the corporation franchise but we will create a new franchise which is the privilege for using the streets of the public. We shall call it a special franchise. We shall say that a special franchise is real estate and you cannot deduct debt from it. All of them have to pay 10, 20 or 80 percent of their income. In some cases street railways went into bankruptcy. We are just about getting around in the state of New York to abolishing the special franchise. It is very difficult to do this because the value of the special franchise in real estate has entered into bond issues. Some of the school districts are supported almost entirely by this.

These are just a few of the problems that arise in connection with the curious system of our country to tax the franchise as property.

Another very difficult problem is the question of banks. That is a very interesting situation. Banks were originally taxed on their capital stock, in fact that is about the only kind of personal property ever reached in the state of New York because when a man owns bank stock {they have to deduct from the man's income}.

I might add that in the very admirable reports that have been issued by the New York Tax Legislators' tax commission, which were prepared by Professor Haig [Hague in original], the recommendation has been made to solve the state and local tax difficulty in the case of railroads and public utilities: we should have what is known as a gross net tax, that is to say, a tax on net receipts, which logically of course, is ideal. It would be inadequate for local purposes in those villages and counties where the expenses grow for the support of schools and building roads. {In case of better times in the disappearance of net proceeds, what is public utilities, and therefore in all cases there should be a tax on gross receipts to be available for this purpose.} By this we get as close to an economically possible point without sacrificing a great deal in particular cases. That is the latest as regards public utilities.

The second class of groups was banks. Individuals were subject to taxation on the bank shares owned by them, but, of course, no assessors found out who owned bank shares. The system developed of making the bank pay on all its stocks and bonds. That system, coupled with the local taxation of real estate, became the prevalent one throughout the country and especially so after the Civil War. Then, when national banks developed, some of these states endeavored to put what was considered to be an illegal tax on national bank stock over against state banks. Congress passed a law saying that national banks would be subject to no more and no less taxation than other capital similarly employed. Under this regulation we have the uniform system of all banks.

In the last few years, new forms of taxation appeared in this country and income taxes were imposed upon banks. State banks wanted to get rid of their obligation and went too far. They brought suit testing the constitutionality of these special taxes on banks. The court held that particular kind of bank [tax] was in conformity with the federal law. The very ingenious and astute legislators now imposed a general tax on all financial capital that you see within the federal law. Capital was now to be taxed equally. It imposed a very high tax on the banks because they had to pay the ordinary tax on all their capital. That was not so bad in itself, but as a result there was caught in this net all the other people who dealt in this, such as private companies. This created a universal consternation that the banks were hurrying over each other to come to terms. {For the last few years, there has been a great fight in Washington to amend the provision to permit in various states that banks pay their fair tax, by subjecting their income tax.} This was almost successful. It would not be long before that clause will be amended greatly so as to permit banks from doing what they did in New York state, claiming exemption from all taxation. {We are in the process of developing a new method in those states where the whole tax will be abandoned. As regards other groups and all groups that tax developed later and started again with the franchise tax on capital stock in Pennsylvania in 80's and in New York little later that seemed to be inadequate, because it left the bonds out.} A good many states have gone over to the taxation of stocks and bonds; that brought about the difficulties of the franchise tax. {You see, if you tax the corporation on property and then tax the value of the franchise also, in order to get away from this, different states have gone over to receipts taxation – some gross receipts and some net receipts.}

The last stage of development is only beginning and it is beginning in New York. You see, we have a tax on land, now we have a tax on corporations, and the claims have gone forth that we should tax other businesses – unincorporated businesses. That recommendation you find in the last reports of the tax commission; they will go through without much doubt. Our corporation tax started as part of the general property tax and has become a business tax. And then we shall have the land tax and business tax side by side with our corporation tax. In certain states we have every survival of the whole system of taxes.

I might add that there is one blot on our system in New York. Whereas we made progress in all the other forms of taxation, including corporation taxation, our general corporation tax applies only to business corporations and does not apply to public utilities, e.g. railroads, gas and electric light companies. We still have a combination of the old and new systems. The reason we have that is to put them all on an even keel and make corporations generally bear their burden. It would in some cases level down and in others level up. Our Electric Power and Gas Companies are enormously over-taxed; on the other hand, railroads are slightly under-taxed. The ideal way means great profit to certain companies and certain sacrifice to others. You would like to contribute for taxation but, on the other hand, when you are threatened with additional taxation, then you will fight to death. Other states have a more equal system. We are in this respect only lagging behind.

You see, then, what the situation becomes and how we tend to have a business tax. When we come to the income tax I will refer to the corporate income tax, and see how that theoretically fits in with what I have said.

There remains one other kind of tax, an abnormal tax levied during the war, now only an historical episode: the excess profits tax. All of you know how these excess profit taxes arose. They arose out of a feeling that during the war, when certain members of the community marched to the front and sacrificed their lives, it was not fair that other parts of the community are not only safe but make more money. For the first time in history we had this movement. The doctrine of what is now called conscription of wealth also arose. We have conscription of men into military service but we did not have conscription of wealth. The theory is that as long as you take a man's life, you can also confiscate his property. Most of the countries took the whole benefit. Denmark and Sweden in 1915 introduced this; Italy followed by taking 50%, Germany introduced it and finally England in 1915 took 50%. In 1917 it reached to 80%. Then a few years after the war it was entirely abolished. In 1916 it was 50% in France and ran up to 80%. In Germany it went up to 100%. In this country, with a special tax on munition manufacturers, we took $12\frac{1}{2}$%, then the following year we introduced a general property tax of 8%; later on, after we entered the war, the tax went up to 20 to 60%. In 1919 our taxes reached their highest point, we took 65% of the profits. Then the tax began to go down and finally disappeared in 1921. In 1917 in Canada it reached to 75%; in Australia, also ran up to 75%.

The theory of the tax was all right so far as it was an attempt to take from the community part of the profit that was due to the war. Practically, however, there was great difficulty. In this country, the opposition was directed at the criteria of what is normal and what are excess profits. You have to measure according to capital. There was a {combination of a tax on earnings compared with capital}, and how are you going to measure the capital? You thus bring those immense difficulties for the last three-quarters of a century: How are you going to evaluate the property, whole profits or loss? Of course, all corporations tried to reduce their capitalization. Even if they did not do that it led in a great many cases to a very undesirable condition, which is, in business practice, called postponing the profits. In the case of the Tea Company, the actual receipts are their dividends. The difficulty of measuring capital is an undesirable system. The tax proved to be ultimately unworkable. It was already so during the war; but as soon as we reached normal conditions, all those who felt that some kind of tax on excess profits would be desirable from social point of view tried to impose a workable system of excess profits taxation. It is all gone and we are not likely to have this until we have another war and then we might have something different. Now, you see, we discussed these taxes *in rem*, semi-personal taxes, land tax and tax on business. Now over against these you have personal taxes:

## 76. **Personal Taxes**

## 77. **Poll Tax**

It is very interesting that governments always oscillate between these two kinds of taxes. When the personal tax was considered bad in France they tried to get the other kind. Now we are back again to personal taxes. We have to attempt to give an estimate of these personal taxes, some of which are disappearing, some of which are becoming more important. Those that are disappearing are: tax on polls, tax on heads, tax on human beings.

Where there is very little private property, that is the only way to squeeze something out of a person, as in Africa, where the poll tax is the only tax. In early Rome it was *capitatia humana*;[191] after a while the property tax developed but it became unequal. In democratic countries you find a combination of poll tax and property tax, but after a while the poll tax begins to show its weaknesses. In some countries they have developed this poll tax according to classes, as the head of a workman is worth more than the head of a beggar and the head of a duke worth more than that of a workman, etc. In England they are always considered, as in later Rome, *nota captivitatis*.[192] So in England you had an uprising in the 14th century. Then the last time poll taxes were employed was in the 17th century. Since then they have never had it. Every country has had a little of its poll taxes at different times. In some countries, they developed class taxes and ultimately income taxes in Germany, and even in India and Russia the poll tax disappeared. A few years ago the only countries who had this were extreme democracies, such as the United States and Switzerland.

In this country, in some states, it has disappeared entirely; in other states, it is kept simply as a kind of political scheme, for instance, in order to get around negro suffrage, etc. In some states, payment of the poll tax is a condition of the suffrage, which leaves the poor class without the vote. In other states, we still keep the poll tax, but as a mere vestige of a former tax, ranging from $1.00 a year to $2.00 or $3.00 or $5.00 for unmarried men. In addition to that, you have a common poll, local poll, so that in some other states it runs to 10, 12, 15 dollars. We find examples of the poll tax, therefore, but the total revenue received is about one quarter of 1% of revenue in this country. Massachusetts is ahead in this; sometimes it amounts to one million dollars. California and Texas are next. However, it is dwindling everywhere and disappearing very fast. The second kind of personal tax is the tax on expenditure:

## 78. **Tax on Expenditure**

The direct tax on expenditures. We find that from the very beginning in Greece; you find it in the Middle Ages, a tax on wigs, silk gowns. In modern times we find these taxes to a very much less extent. In England, when they were anxious to get revenue, these taxes were on expenses, carriages, servants, horses, dogs, watches, and a few others. In France they have them, also in Germany, such as the tax on nightingales.

In this country we have the tax on carriages. This was once a great constitutional problem; they were deciding what was a direct and what an indirect tax. It was abolished in 1802, reintroduced during the war with England, {and not only then but they introduced also a few others, such as taxes on household furnitures, billiard tables, coal, gas, etc.} In 1872 it was repealed; then, of course, during the war we had taxes on pianos, furs, etc.; these were known as nuisance taxes during the last war. We also had had it during the war with Spain. So you may say that taxes of this kind are always used in emergencies and rightly so. Inasmuch as we know that expenditure is not the best criterion, here, too, we have a dwindling of personal taxes. Now we come to the general property tax, but this also is dwindling.

## 79. **General Property Tax**

The general property taxes are discussed under four heads:

(1) History
(2) The facts
(3) Defects
(4) Reasons for their disappearance

History: In mediaeval history you find the same thing. There are always four stages in the history of the property tax: (a) Taxation levied upon property in land. (b) The addition of a particular piece of personal property, first tangible and then intangible. (c) Summing up the amalgamation of these three, a general tax on men's property. (d) The beginning of a solution as first intangible and then tangible properties are deleted from the list, so that finally nothing is left and the general property tax disappears. That is the history, always.

The general property tax was developed in Germany and in other parts of Europe in the 13th and 14th centuries and decayed in other parts of Europe as a result of economic development. By the 18th century, it had disappeared everywhere. Now, in this country, it is curious that we have both the most developed and least developed economic conditions; accordingly, we cannot generalize the development of the tax in this country. We can say that it existed by 1789 in a few of our states, and by 1825 in most of our states, and that for the last 75 years our general property tax was both introduced in the constitution and finally gotten rid of.

In our discussion of the general property tax, we saw how the property tax grew up in this country. Individual ownership of easily assessed pieces of land was among the favorable conditions; from the time of the colonial period in the 18th century the property tax gradually developed. When problems arose in France and it became necessary to tax property, a great investigation was undertaken; and in the reports you find a very interesting point developed. In the Southern states the economic situation precluded the possibility of any kind of property tax and they did not come about until later. You may say that by the middle of the 19th century the general property tax had become universal. Shortly afterward, it began to break down. I remember when, in the 80's and 90's, I first began to write on these subjects as they were found in New York and Massachusetts, I was very severely attacked. The state tax commission still believed in the general property tax. But by 1900 you find much weakening. In 1910 they were all pretty near gone. By 1920 there was not a single state left in which the state tax commission and others still defended the general property tax in principle.

It is not necessary now to review what was necessary a few decades ago and call your attention to its defects. When we compare the capital values and the income criterion of ability to pay, what was true about half a century ago is no longer true today, and then in practice, of course, the fundamental difficulty has arisen from the impossibility of getting revenue from the intangible property tax. There is more of this property than we have in real estate. Of course, land is the foundation but it is not social wealth. There is the difficulty from the point of view of social wealth. If you borrow money and if you mortgage your property, the mortgage in the hands of somebody represents individual wealth, because it is a source of revenue and {therefore when we say that on the basis of real estate built up on the structure of individual wealth, we are referring to individual wealth rather than social wealth.} Now, the tax having broken down, we would expect to see it disappear first in the stages of most developed economic countries. After the general property tax became universal it was entered into the constitution and a provision was made in the constitution. Europe did not have this. But with us, in almost all states, provision was made that all property should be taxed uniformly. It so happened that a few states escaped; this led to a very embarrassing situation. At all events, where you have constitutional prohibition, it is much more difficult to break up a general property tax and go back again to taxation of particular pieces of specific objects because as soon as you tax specific objects you are opening the door for discussion by putting different rates upon different classes. In New York state there was no such thing and at the beginning of the 80's we started specific taxes on certain kinds of property, railroads, etc. That was an entering wedge. Then in New York, for instance, about a decade ago the tax on intangible personalty was completely abolished and placed on something else.

There was a movement in the last five years to enact in law what has already by now become a fact: to abolish tangible personalty taxation, which does not exist in a state of any importance – and one from which we get only about one million dollars. The tangible personalty tax has disappeared. Other states have not gotten so far as New York, though a number of states have gone so far and abolished taxes on intangible personalty, Massachusetts for instance. Some other states are still in the earlier stages of trying to levy taxes on specific objects rather than on general property.

That is the constitutional struggle in this country. In going into detail, of course, it is very plain [plane in original] as to what is happening in this country today in proportion to economic progress: we are getting closer and closer to the day of the disappearance of the general property tax. About the only place in which it would be left is in the country district of Switzerland. We pass it by not because it is not important from the point of view of revenue – it is still the most important tax we have in this country – but the defects of the theory need no longer be discussed because it is on the decline.

There is, however, one other aspect which came in very recently – the capital levy.

## 80. **Capital Levy and Capital Increment Tax**

The capital levy is simply the British way of putting the general property tax. It is a more accurate term in economics. We are contrasting this not with property but with capital. In this country we mean the capital value of property whereas in England it is something new, so they call it the capital levy. And, inasmuch as the property tax is not found in Europe, there has also been introduced the difference between capital levy and the general property tax. The general property tax is an annual tax whereas the capital levy is a tax levied on property once and for all and at a rate of practically 2 or 3%. Whereas capital value gave rise to an entirely different problem, not to pay the current expenses of government, as our general property tax does, but in order to pay off the debt and therefore diminish the capital sum. [The foregoing is somewhat convoluted: The property tax paid for current governmental expenses; the capital levy was used to pay off government debt – or to pay extraordinary (wartime) expenses.] Of course, you need a higher tax to pay off a large debt. That means the tax rate will run to 30, 40, 50, or 60% of a man's property. After the embarrassments of the war, this was a very natural phenomenon.

The first time a capital levy was ever suggested was shortly after the public debt became important during the war with Great Britain, thereafter the war with Spain. From 1717 on you find that we have writers on the subject of the capital levy. But soon people forgot about it. About a century later no less a thinker than Ricardo advanced the same idea, after the war with Napoleon. Ricardo was a great protagonist of a capital levy. Of course, he said, there was no chance to introduce it. Great Britain in 1914, as in 1815, did not pay all her debt. Debt was then under a million pounds, now it is about 48 billion dollars. {When taxes have already been screwed up to the point where they compelled everybody to pay taxes, if small incomes did not pay, but then small incomes later on also paid 30 or 40%.} In other countries, such as Italy, Germany, Romania and Czechoslovakia, the situation was worse.

There has been a great literature on the capital levy. We never needed this tax in this country because we made so much money out of the war before we entered the war and we came out almost even. {When we came to close quarters with it.} In Italy and Germany, they found practical difficulty. {What was virtually changeable, it was payable in installments and made an addition to the entire tax. Even that did not work and further payments were done away with.} In only two places in the world has the capital levy been carried through, Czechoslovakia and Romania. They had to have money to pay off the debt, they rather needed money. What they did in Czechoslovakia was to pass a law on Sunday night calling for pay[ment on] pay Monday. The law provided for every bank to pay 20 or 25 percent of all their assets. Of course, a rude [crude?] method of that kind will not have any chance in the more developed countries. People were already stung in those countries by the issue of paper money and to lose a little more did not create much difficulty.

Now in England it is still on the program of the Labor Party but even after they come back to power again they might not put it through on account of its many practical difficulties. Land value taxes in England went to pieces simply because of the difficulty of determining the capital value of land under the complicated tendencies. And, if land value did not work in England, how much greater will it be to determine the value of intangible property? So the capital levy represents an absurd aftermath of war, and may be dismissed as an historical fact.

## 81. **Personal Taxation**

Personal taxation at the present time is not on the decline but on the ascendence. Poll taxes have gone, the general property tax is almost gone, and the income tax is getting much more important. We will discuss it under three heads: (1) History. (2) Facts. (3) Problems.

(1) History: This is divided into five stages:

(a) Mediaeval stage, where you find the income tax developed in Florence, when the economic conditions were more like those now in this country, but it disappeared after the disappearance of industrial democracy and the conquest [i.e. ascent] of feudalism in Italy.

(b) Sporadic stage. The emergence of the tax as a war measure. Of course, most taxes, you remember, start as war measures anyway. There were income taxes in 1799, when Pitt[193] tried to make it carry on. It started with a tax on expenditure, and finally became an income tax. When the war was over in 1816, the tax was repealed and Parliament ordered that every piece of paper in reference to taxes ought to be burned. The same thing happened in this country in the war with England. Had the war lasted longer, we would have had an income tax then, because the committee was trying to put the income tax through. This committee was composed of the framers of the Constitution. Peace was soon declared and it ended that. When civil war broke out, among other taxes the income tax was introduced. This caused a great consternation and when the war was over it finally disappeared. Everybody at that time felt {that tax was much more productive than the revenue}. I remember when I was a boy in 1877 that one of the songs was "Sammy, Sammy, [Sami Sami in original] pay your tax". When the tax disappeared, everybody thanked God.

(c) The modern beginnings. That began in Great Britain in 1842, when Peel[194] wanted to do away with the Corn Laws. He found that they could not sacrifice all that revenue; finally, they decided to have the income tax but only temporarily. But it turned out differently; every two or three years they had to renew it. Finally, in 1870 came the real fight. It is noted that Gladstone[195] said, if you don't vote for me you are going to have an income tax. But it is only since the 90's that the income tax in Great Britain became very important, and of course during the war. The survivals of this outside of Great Britain you find in Switzerland, where a certain kind of income tax was introduced to round out the general property tax. When Italy became a great united state in the 60's, the liberals, being very much influenced by England, adopted the English system and in 1860 the income tax was adopted in Italy.

(d) This stage is [that of making the income tax] a permanent part of the system. Japan in 1877 adopted it as a permanent measure, and some of the German States, e.g. Prussia, followed suit. Holland followed three years later. In the meantime, Australia and New Zealand began to introduce it in a gentle way in some their states. Finally, the democratic movement had reached [sufficiently far] that in 1894 we passed the income tax, although the way in which it was levied was declared unconstitutional, as, according to our constitution, direct taxes ought to be proportional.

(e) This is the stage where the income tax begins to become the most important, overwhelming element of the existing tax system. Even in France, a great effort had been made to prevent this. Only in 1917 was it possible to introduce the income tax in France. There was great opposition and it was referred to as a communistic idea, etc. {In the meantime, the movement had been so strong in this country that it was left to the decision of the Supreme Court, and the great social advancements that have been made in England, Germany, Australia, so that this issue led to a great agitation and income tax was introduced in 1913.} The feeling is that in these earlier times it had to be classified according to the war conditions. In England and Ireland, where the income tax was levied, one Irish gentleman included this notice with his income tax form: "Gentlemen, take notice, I have cut the throat of my cats and shot my dogs. I have dismissed all my servants except my wife so that I am not liable for any tax whatever". You see how ideas change.

(2) In regard to the facts, we must be very concise. There are at present three fundamentally different {taxables} of the income tax. The first and most important is the oldest, which is originally an English tax. England levied it in 1799. Because some people declined [to pay] the entire income tax, it broke down completely. There was no way of checking up, because no one thought that men would come to your house and look through your books. If they had tried to introduce such a thing as that in France, there would have been a

revolution. Therefore, they devised a new method, a method of schedules, which divided a man's income, on each part of the land, management of land, etc., and introduced important bills providing that the tax should be paid not by the men who received the income but at the source of the receipts. For instance, in England, of course, all land is listed in order to get to the owner. The tenant who paid the rent had to deduct from his rent the amount of the tax paid to the government. The same thing was true all the way through.

In the United States the income tax was recommended during the War of 1812. It was utilized for two years in the Civil War. It was again enacted in 1894. During the interlude [after the Supreme Court decision ruling it unconstitutional] the constitution was amended, in 1911, and it was introduced in 1913. Since then it has become our national system. It was not until 1911 that the movement started in the states. Wisconsin made the important first step and New York and Massachusetts followed two years later. Now it has spread to 12 or 13 states and it is spreading to the others. In other countries, the situation is very much the same.

The German states, especially Prussia, started with the income tax in the 90s, or about a generation ago. Then, of course, during the war its use was very notably accentuated when the systems were taken over by the German federal government. In France the income tax was the last to be considered, owing to the prejudices still felt against all personal taxes. Ever since the Revolution, the system in France throughout the 19th century was quasi-personal. But finally the law was put into operation there during the war in 1917, and since then it has attained very great proportions. In Italy it started a little earlier, back in the 60s. The income tax is a more important part [phase in the original] in their system. In other countries, such as Switzerland and Austria, the income tax came in at the end of the century and, especially due to the burdens of war, we may say that the income tax has been introduced all over the world. Now, then, that is the history of the income tax.

We will say a few words regarding the facts: Here it may be said that there are three types: You have, first, the English system, where you divide the tax up into schedules and attempt to levy the tax by making personal payments, rather than personal receipts, called stoppage at source. That, of course, was the only system in England until two decades ago, when it became necessary to introduce a graduated and progressive system. If you have this system, you cannot apply the scheduled one. You have to proportion the tax by a man's entire income. They introduced the super-tax, as they call it, which is calculated on a man's entire income. It has worked very well in Great Britain over against the preceding system.

In Germany and the United States the chief application of the other system is a lump sum tax where the tax is imposed upon a man's income as a whole. It is true that we differ categorically; we don't call it a schedule but they are simply added together into tax, then it is imposed upon both, no difference in rates or in methods.

The third system is typified now by France, Italy and by some of the other countries that have been influenced by France, and that is a curious combination. France, you remember, all through the 19th century, had its so-called *impot rèelle*[196] [reel in original] tax all things and there is probably a little misapprehension of the English system of schedule. They called their new system the schedular taxes. What they did, was that they took the whole thing – land, business etc. – and injected to these a little bit more of a personal element by making allowance for a minimum of subsistence. But in reality still it can be very largely what we might call a tax on source of income. So that they have the basis of their income tax upon different kinds of income – a different rate and different methods. And, then, on top of that they have a general income tax graduated according to a progressive system. So that, in France the income tax is composed of two parts: scheduled taxes – income from land and business and on top of that *impot global*[197] [*impot globule* in original] – and the tax on the entire income [awkward structure]. They misapprehended the English system, because, as I pointed out, that is the schedules used for administrative purposes.

At all events, you have these three different systems of income tax. The stoppage-at-source system, primarily in England; the lump sum, such as in Germany and the United States; and then this combination, schedule taxes, in some of the Romanic countries. Of course, you can say the French and Italian system simply marks a transition from the stage of the semi-personal tax toward the personal tax and that they haven't quite reached the stage which has been reached now in Germany and in this country.

I will not go into detail, which you can get from any book, except that for the last few years the rates have been going up enormously. In England, the rate went up to about 60%, now the maximum rate is 50%. In France, schedule taxes ran up to 10 to 15%, the general tax ran up to 60%. So that France has the highest rate. Germany also is high. The Italian system is noteworthy for being divided into schedules and different groups, taxed at very greatly varying rates and introducing far higher rates than anywhere else in the world. The principle of differentiation under the new law is such that the rates vary from 5 to 45%, not on the amount of income but on the kind of income. For instance, we [they?] do not tax all dividends and have very low rates, but the tax on land runs up to very high rates.

For this country, it might be worthwhile to take the facts down. These are not minimum and maximum rates. In 1913 the rate was 1 to 7%; 1916, from 2 to 15%; 1917, from 4 to 67%; and after we really got into the war in 1918, from 6 to 77%. Then came the reduction. In 1921, from 4 to 73%; in 1922, from 4 to 58%; in 1924, from 2 to 44%; in 1926, from $1\frac{1}{2}$ to 23%, which is our present rate. It is probable that it will be reduced somewhat. Now those are the facts. The revenue, of course, can be enormous. The revenue of this country amounts to $2\frac{1}{2}$ billions of dollars, and in all other countries all the expenditures are being derived from the income tax.

(3) What are the chief problems? The first problem is, what do you mean by income? That I will pass over now. We will discuss it later. All I want to point out is that I would like to call to your minds that in this country we still in part blame methods which have been either abandoned by other countries or not even accepted by them. For instance, we limit our money income to psychic income, house rent, what farmers spend themselves. In the second place, we still include the concept of capital gains in income. This has not been abandoned in Germany and never accepted in England, as seen in the report of the special committee of the House and Senate. In our system, the tax is only $12\frac{1}{2}$%, and the committee defends the system largely because we need the revenue. The Secretary of Treasury came out with the recommendation that we abandon the whole plan. You will either have taxes on capital gains as such or do as the English do, also the Australians and in other countries, make a distinction between capital and income. It seems rather difficult to defend a high grade system. As regards the other point, {recourse to irregular income to introduce the gift tax}, that has been abandoned and, on the other hand, inheritance is a separate tax. As to what is income, great progress has been made in this country through the ruling of different authorities on questions of depletion, administration and so on.

We will now take up the question of the rates. Here, you see, we have, first, the question of exemption; second, graduation; and third, differentiation. As regards exemption, we are gradually raising the amount of the exemption, so as to make our minimum not simply a minimum of subsistence but a minimum of comfortable subsistence. Other countries haven't done this. On the other hand, in England they provide for dependents to a much larger extent than we do. And the movement is only beginning to permit exemption for other things, such as insurance premiums, illness, etc. We have only the beginning of the net income idea in this country.

The second question is graduation. That need not be differentiated in this country. The only point is, how high should graduation go? We have under this head: (a) Need for revenue, and (b) The defects of administration and the chances of evasion. In a country like the United States, the older views with reference to graduation have been seen to be largely inadequate in the earlier period of very small graduation. In Germany, Switzerland, and in Great Britain, the great mass of the tax comes from the lowest classes, and highly rich incomes really yield insignificant sums. But in the United States it turned out very differently. With our great fortunes, that developed incomes of from one to three millions of dollars a year, the totals are quite considerable. You will find that these higher classes contribute by no means a small share of the income-tax revenue. The maximum rates have been reduced, however, from year to year according to our needs, and we have come finally to the lowest. It is not likely that we shall go any further down. They are nevertheless very much higher than before the war. In England and France, rates are very much higher than in the United States.

The third question is differentiation. Here, after a great many struggles, we finally adapted differentiation between earned and unearned incomes. In England it is limited to comparatively small sums, 10 to 20 thousand dollars. We went up to 20 thousand dollars. This is probably too low here and it may run up beyond that. The reason why it has not gone further is that, when you get beyond a certain limit, unearned income is characteristic of wealthier people; but when you get to a very large income, like the salaries of railroad presidents, the salaries of a few lawyers or doctors, which run to some hundreds of thousands of dollars, people have not much sympathy – for instance, for a lawyer charging $1000 a day.

Third, administrative methods are still very important. Our administration is far inferior to the British and although better than the French may be also better than the German. The administrative methods in this country suffer largely because of our extreme democracy. The difficulty with all democratic governments is that they do not pay sufficient amounts to the experts. In England they pay more. Private corporations pay much more than the government. The new law has greatly increased the salaries of our income tax revenue officials, yet, if you increase them too much, you have to change salary structure [change all the graduation in the original], because you cannot pay an income tax lawyer much more than the members of the Supreme Court. These are the difficulties which we have to cope with. In some respects, however, we are far in advance of other countries and that is in the readiness with which the individual adjusts himself to the law. Some of our administrative methods, if tried in France, would lead to a revolution.

We have now become complacent and, as a result of that, the evasion of income tax is much less than it used to be, far less in Italy, probably not so great in England. In this country it is impossible to figure how much is evaded. In

England, the estimate is 10%. The chances are that evasion is more pronounced in the middle of the schedule. Almost every wealthy individual has their income tax return made for them by accountants and the probability is that there is comparatively little evasion. But in the cases of the new-born rich, all those incomes which change from year to year, and also small businessmen, there is an immense amount of fraud which seems to be inevitable. But, on the other hand, the income tax is so successful from the point of view of the reaction of the individual and public opinion, it is not considered as inequitable.

The problems under the head of administration are: (a) Self-assessment or exercise of examination. We have a combination. (b) Shall the tax payment be a lump sum or stoppage at source – which was tried here and did not seem to accord with our American methods.[198] Our system is that of information at source. (c) Individual or family return. In England, rates are very much higher than they seem compared with this country. In England, you must make a joint return. In England not only the rates, but the actual rates, are very much higher, and when we compare it is very different, it is a separate return and not a joint return. (d) Ought the income tax be paid on annual or one's average income? This is important in Australia. Take three years. The first year, income is $1000; second year, $2000; third year, a loss of $30,000. The system in Great Britain was to take the average. Our system was to take the annual income [return in original] with certain modification; certain attention was paid to this before England abandoned this system. On the whole, the arguments in favor of the modified average system seem to be the strongest even for countries which are not exposed to certain oscillation, and the annual system seems to be a little difficult in that it does not make allowance for loss. If a man loses one year and gains the next year, government does not make allowance for the loss. The ideal system would be over a period of time.

The fourth problem is that of the subject of the tax: Whom do you tax? The individual?, of course, and also the corporation. That has become a topic which has not been thought about yet and on which a great deal of work has to be done. In some countries you tax the individual, for instance, the stockholders in a corporation. In some countries, the corporation pays the tax on income but deducts it from the dividends, so that in reality the stockholder pays. In some other countries, you have a different system where the individual pays and the corporation pays also, like our system in New York State; the individual and the corporation both pay. In other countries, you have the type which we find now in our federal tax. You have a tax on the corporation and you have a tax on the individual, but you allow the individual to {deduct from his income his dividends derived from corporation by our federal government,} you allow that only for the purpose of the normal tax, but not for the purposes of all super-tax. You can defend the first system, you can defend the second system, but I don't see how you can defend our system.

Now, as a matter of theory, I recommend the second system: The English system looks upon income as such, only so far as the individual is concerned, whereas our system says that in addition to the tax on the individual as such you can justify a tax on the business as such, just as you have a quasi-personal view. The ideal system, it seems to me, is the one towards which the world is moving, where you have an {income tax upon individuals as a personal tax upon corporations} and [tax] other businesses as a tax upon business.

In closing, we spoke about different systems of taxation [plans in the original] followed throughout the world. Now I would say that this is by no means settled in either theory or practice. I think that the last word said in New York State is correct, that is to say, we should look upon the personal income tax as only a part of a larger system. In that we must say, by the way, that all these personal taxes, the so-called quasi-personal taxes of which the best example is in New York State, are taxes on land. A second example is coming to other countries, a tax on business. Then we shall have a logical and entirely dependable system. Between them we should have this combination of taxes which would associate all the important elements because the only other tax on things, as against personal [taxes], would be a tax on capital as such, over against land. We have seen that in discussing the personal property tax, all that is variable by taxing [taxable in the original] business; and, outside of capital invested in a business, you have only the other forms of personalty which ought not to be taxed at all, either because they represent consumption rather than production or because the capital takes the form of such intangible personalty that it is impossible for assessment. If we supplement our personal tax with a tax on things, land on the one hand and business on the other, then you are introducing the modern element of wealth. We are on our way toward that in this country in the federal government. In England they are not on the way because this concept of impersonal taxes side by side with personal taxes has never reached England. They don't realize their local returns and have gone beyond the income tax. We have been beyond that for a long time, because the personal income tax was constitutional. We are prepared for the next step.

The only other question now left is the question of double taxation by the income tax, of which we spoke in discussing double taxation. If we have this business income tax side by side with the personal tax, you solve the entire problem; the personal tax is always levied on the individual according to his residence and the corporate income tax or business income tax is naturally levied in the state in which that business is carried on. If we follow that plan, we will move toward the solution of the entire problem.

That is really about as much time we have to talk about the income tax. The income tax is the most important tax in the world today and soon will be the most important tax in our state. It is not only significant from the pecuniary point of view, it also is receiving closer study in our jurisprudence; it is only beginning to attract attention. Let us leave the income tax and say a few words about the inheritance tax.

## 82. **Inheritance Tax**

What we call the inheritance tax, in England they call the death duty. It differs, on the one hand, from the personal tax like the income tax and, on the other, from the impersonal and quasi-personal tax. The reason is that the tax is of two kinds, a tax upon either the individual receiving the estate or the estate itself; the latter is the impersonal tax. Because of that fact I will call it a mixed tax. Now, then, I will discuss this under four heads: History; Facts; Problems; and Theory.

1. History. We find this tax far back in antiquity, in some of the Oriental countries, and in Rome at the time of the Empire; we find it used largely for war purposes. In the Middle Ages it was in the form of feudal relief sometimes known as *heriot* in England. In France it had different names. "Insinuation" they called it, why, it is hard to say. A curious name was "last penny". But those are all connected with the feudal method. At the beginning of modern times, the point of view was rather different from the fiscal point of view. The earliest beginning you find at the end of the 16th century and the beginning of the 17th century in some of the Dutch towns. In England it began in 1694. A century later the probate duty appeared and in the middle of 19th century that applied to real estate and personalty in general. As a result of the French Revolution, they introduced the idea of a collateral tax for distant relatives. The same idea is found in a few of the European countries, in fact, we introduced the probate duties in this country also.

Now the most recent development is due to the democratic movement, the idea of taxing wealth just as does the income tax. The tax started in Switzerland in 1848 and was the only example until the end of the century when the movement for the inheritance tax began in Australia and New Zealand in 1890. In Great Britain it started in 1894; the death duties were generalized, then a progressive system was introduced from Australia. In France it was introduced in 1901; in Germany, for collaterals in 1904, and in Japan in 1905. With us it is always used in connection with the federal government in time of war. It was used in 1812, the Civil War, the Spanish war and finally in the last war, in 1916.

In the meantime, the movement had spread to our states and took two forms, of which we had a few sporadic examples. The movement now developed in the states, {first for collaterals, and then taken in New York in 1891.} The movement was given great impetus by the decisions of the Supreme Court that the inheritance tax and the progressive federal tax were constitutional, the Court holding that proscription of uniformity meant only [all in original] geographical uniformity. They reserved, however, the right in particular cases to find confiscation, etc. The history of the inheritance tax, you see, was the progress of democratic ideas. Of course, it was primarily correct, so that the rate increased and today inheritance and death duties sometimes go side by side with others in taxing wealth.

I shall say a few words as to the facts: In England, death duties are composed of three parts at present: (a) Estate duty – a tax on the estate as a whole. Since 1894 and rates have gone up gradually until 1919 when they reached the maximum figure, rates going up to 40% on 2 millions of dollars. [b] Side by side with this estate duty you have the legacy duty, which is a tax on personal property going to collateral heirs. That runs up to 10%. [c] The succession duties, which apply to shares received by individuals in their legacies. So that the tax, instead of being 40%, amounts to 50% and 60%. In France, especially since the war, although it began in 1901, they had three or four great additions, one was in 1910, another was in 191[7?] and again recently. In France the rates are the highest in the world and they are two parts tax on the share which they call *mutation*[199] [?] or, as we say, transfer tax, and there is the state tax, *capital net globale* [?] – entire amount. The tax on the one runs up to about 40% and the other one runs to 60-odd percent. If a distant relative received, it would amount to 100%. The law, however, says that in no case would the tax run more than 90%. It is not known, however, if there was any case in which the entire thing would go to the state. In Germany also now the taxes are very high. In 1925 the maximum ran to 60%.

In this country, you remember, we had quite a progression. At first we had the federal law, which in 1916 was imposed only as an estate tax, and that started a progressive running up, first to 10%; then in 1917 it was increased to 20%, in 1919 to 25%, and in 1924 to 40%; then in 1926 it was reduced to 25%, at which level it now stands. The state laws have undergone a very great change. Almost everywhere after 1900 more and more states have added the tax, until now every state has it, except two states who, in order to attract people, give freedom from the inheritance tax – Nevada and Florida. Direct taxes and collaterals ran up to pretty high figures, the collateral ran up to 30%

and in the case of direct heirs ran to 5%, 8%, and 10%. You have to add the federal tax to the state tax, and if you add, you find that, in some of our states, rates are as high as in England and Germany.

One recent great development made the federal tax, state tax and commonwealth taxes shared taxes. This owed to the agitation to have the federal tax discontinued and have the inheritance tax left only for the states. In order to counter that movement, recent laws provide that 80% of the federal tax should be returned to those states which levy the similar tax. States have now been tumbling over each other to get that tax and some of the states have both new ones and old ones, though some states have given one or the other up. So revenue from one diminishes and the revenue from the other increases. You find this in Australia, Switzerland and Canada etc.

The chief problem with the modern inheritance tax is the question of rates. Shall it be a proportional tax or shall it be a graduated tax, and with what exemptions? Owing to a difficulty, in this country we worked out a peculiar system in order to develop an inheritance tax. Here it started, you see, only as a collateral tax. Then, as it was to apply only to direct personalty, the rest was left out. Finally, it was realty to which it applied. In the course of that development, in order to make the tax agreeable to the public, the argument offered by the widow was met by having a very high exemption. In most of our states you don't tax unless it is more than $50,000 to $100,000, whereas in Europe there is no exemption. That is the first thing to remember. Our tax does not apply to small estates. The next movement involved progression. When graduation was introduced it was introduced very gradually. Today graduation is very slight and it is found only in the federal tax. The argument in favor of progression is just as strong in the inheritance tax as it is in the income tax. We may say that now it is accepted virtually everywhere.

As to both graduation according to collateral and the second question, namely, the distinction between a direct and collateral tax, it everywhere begins as a collateral tax. The argument for graduation in the case of a collateral tax is different from the argument in the case of a direct tax. In the case of a collateral tax, the idea of family comes in, but in the case of a direct tax it is a matter rather of the ability to pay which is considered.

The third question is whether to put tax on the entire estate or on a share. According to the original theory, which was ability to pay, inheritance is considered to be a supplementary income tax and, of course, the argument would be very strong in favor of having the tax as a share tax. In the New York report there was a very elaborate argument in favor of a share tax over against an estate tax. The success of the estate tax is due to two reasons: (a) There is more of an endeavor to restrict large fortunes by making them pay their share.

(b) The administration is much simpler. So we now have only the estate tax and it is more and more getting to be the system throughout the world. The conclusion would be that the best system is a combination of those of both England and France. By having both you meet both arguments provided the joint rate is not excessive. Of course, in France both rates are entirely too high.

The fourth point is connected with frequency. A gentleman, an English duke died, and his successor had to pay a very large tax. His son was 70 years of age; he passed away and the estate had to pay another tax. His son was about 48 and was killed during the war, and within a few weeks his son also died and left a baby of six years. Very little of the estate was left because about 30% was taken away four times. In South America, Chile passed a law that the same property should not be taxed again within a certain number of years. In England, they passed a law to prevent a levy in case of quick succession.

The fifth problem is avoidance of double taxation. They already allow deduction in double taxation for the same person.

The sixth problem is administration. Should a tax be collected by government authorities? In this state, for instance, government authorities collect the tax with considerable success. The other side of the question is the danger of evasion. It is very easy to evade the inheritance tax. You cannot avoid the inheritance tax by incorporation, however. The most important way is by making trusts, which is done by rich people who give away all their property before death, the heirs reserving for themselves and the wife something to live on – a modest income of perhaps a few millions of dollars a year. If it is in connection with gifts, that has not been as successful as in Europe.

Another question is whether the tax should be a state or federal tax, a typical American question, and a political question. {With the interference with the fiduciary change, which is not at all improbable, we had great many arguments.} We had people like Carnegie[200] who wants never to impose an income tax; he wants to impose all taxes on inheritance. Mr. Mellon,[201] on the other hand, says that you ought not to have an inheritance tax at all. It is remarkable that the opinions of members of the wealthy classes should differ so much. In England, these things are not discussed any more, they don't discuss the income tax, neither do they discuss the inheritance tax. {But, we have federal tax developed and as they have federal tax disappeared turn their attention upon state affairs when different political fight is on.} The great majority have accepted the inheritance tax. Of course, workers and farmers would naturally be in favor of the inheritance tax. One argument raised is that the federal government is encroaching upon state rights. Its history is unsound, because the first tax of importance was the federal tax and the first kind of tax discussed was the federal tax at the time of the war with England. At that time the framers of our Constitution did not think that inheritance rights would be left to the states. We shall have either a federal or state tax; there is no probability of a combination.[202]

In this country today we get very much less in revenue from the tax than in other countries. In England they get several hundred millions from their death duties. In this country, with 125 millions of people, the federal and state taxes bring in somewhat less than England, whereas the United States is about three times as populous and three times as wealthy as England. So that if we had the same proportion our inheritance tax would bring in a billion dollars. Of course, we do not need this. However, we could reduce our income tax to half. It is not likely that this will happen.

I shall say a word or two more as to details you will find in the *Essays in Taxation*. The main arguments are the limitation of inheritance argument, which starts with Bentham, the other side of which is the charge to be considered as a kind of cost of service or value of service, and is Mr. Carnegie's idea; and the other is the tax idea, either an indirect tax – a tax on transfer – or a direct tax of some kind. Of these arguments the first applies clearly only to the collateral inheritance tax. In modern times we do take a large portion where family connection is weakest. The fee idea can be eliminated because we have gotten beyond that; also the idea of indirect tax, which is a legal argument. In this state we call it a transfer tax; in France, they call them taxes on transaction.

We come now to the other side of it. We now understand that it is sometimes a tax on the estate as a whole and sometimes a tax on a share. There are two arguments. The accidental income theory can be invoked for the inheritance tax on a share. If the income tax is the best tax of ability to pay and if it includes a regular tax, you see inheritance as coming accidentally and the inheritance tax may be considered to be a supplement to the income tax. That argument would not apply to the more modern form of income tax as an estate [a state in the original] tax. If you look upon it as an estate [a state in the original] tax, you can consider it to be a tax on wealth supplementary to either the tax on wealth as property or the tax on wealth as income. {It could be defined here as an additional tax making something over and above or as supplementary tax is imperfect any way.} Governments have gone ahead, they always have, and as we know the inheritance tax in its modern form has been simply a result of the democratic movement and the endeavor to reach the larger aggregations of wealth theoretically, I should say you could [thus] define it. Of course, if you are socialistically inclined, you would say to limit fortunes, but very few go as far as that.

# Part 2. Indirect Taxes: Taxes on Exchange and Consumption

## 83. General Considerations

I might point out that there is a great distinction between Roman and Teutonic countries. Romanic countries are closer to the mediaeval system, where indirect taxes are more common and you find them considerably more than in Anglo-Saxon countries. After all, the entire revenue derived from this source is not important and is gradually dwindling in Romanic countries. The reverse is the case in Anglo-Saxon countries. Entirely apart from prohibition, the tendency is to levy a tax on wealth rather than on consumption. As to the relative merits, some people will agree with the opinion of Gladstone, who, in his speech on the Budget, declared that he was in favor of direct taxes. However, if he spoke today, he would speak differently. We make use of indirect taxes in cases of emergency, such as during the war, although, of course, this by no means implies that you can get by with only the system of direct taxation in normal times. The arguments could be put as follows:

The advantages of the indirect taxes are:

(1) They diminish the burden on the direct taxes.
(2) The psychological reason. It is least felt.
(3) You reach revenues which otherwise perhaps might not be taxed at all. Everybody makes some expenditure. If a man has all his investment in the United States in bonds, that would not be reached at all.
(4) That the system {especially in some of our bonds} is more productive and more elastic. You increase the rate of tax up to the certain point. The matter of elasticity, however, should not be over-stressed.

Certain other advantages that are often mentioned, however, are not true. The diffusion theory is spoken of. Of course, we know from our theory of incidence that in all indirect taxes it differs. The second is the psychological argument. The third argument is not important, inasmuch as it is built upon an argument which is not true.

On the other hand, the disadvantages of indirect taxes are as follows:

(1) In order to get revenue, this tax is apt to disturb industry. Look at the tax we have on automobiles in the last few years. I mean the federal tax upon manufacturers of automobiles, also other commodities.
(2) That the indirect taxes are apt to bring about artificial changes and that all will favor wealthy members of the class.

(3) That expenses of collection are high. They are apt to be higher than any direct taxes. For instance, when we had the whiskey tax, the cost of collection in North Carolina and also in Virginia sometimes reached to 30%, and even under more favorable conditions you find, in Europe and in France, especially, it costs 5%, 7%; in England a little less.
(4) Inequality of the tax, because of the failure to distinguish between necessities, comforts and luxuries. That is the main reason why, beginning from the Middle Ages, tax reformers wanted to have necessities exempted while in modern times the democratic movement has resulted in the disappearance of these taxes.

There were great arguments over the sales tax. One of the finer objections is the fact that they really interfere unduly with individuals. There is not very much difference between the way in which our income tax is administered and the way the soft drink tax was administered a few years ago. The nuisance taxes everybody considered more or less unconstitutional. In general, therefore, all you can say is that in the more democratic communities the tendency is to restrict indirect taxes for normal purposes to the lowest possible limits. We shall see in a moment that the whole tendency is to restrict indirect taxation to the smallest possible number of commodities that yield the greatest revenues, taking very little expense for collection and out of which you get a great deal of revenue.

## 84. **Taxes on Commodities**

Historically they developed out of what came to be known as excess taxes. It was the Dutch who developed them, and when William and Mary came over to England they brought the tax to England. In this country, in New England, you had the property tax; in the South, the poll tax; in the Middle Colonies, the excise tax. There is a great difference between Long Island and New York, as New York was first settled by the Dutch, but Long Island by New Englanders. This excise system had also been more or less known in different forms in France and pretty much all kinds of commodities were taxed. That is one of the reasons why excise taxes are so unpopular in England, where they had developed considerably. The excise tax system in the 17th century and the beginning of the 18th century started with taxes on drinks, some food and then especially on certain other commodities. During the 18th century more money was needed. The list of taxed commodities increased very greatly and in the beginning of the 19th century almost everything you could think of was taxed. You remember when, in this connection, Sydney Smith[203] wrote those caustic letters on the repudiation of the debts of states. However, one by one these excise taxes gradually were taken off until today they are limited to a very few articles, such as whiskey, tobacco, etc. The same thing is more or less true on the continent, although their process of simplification has not gone so far.

In this country the excise tax began under Hamilton's régime. When he needed more money he put taxes upon sugar, snuff, etc. The Republicans soon came to power and abolished it. During the War of 1812, the tax increased on iron mills, hats, etc. But it was abolished in 1817 and reintroduced in the Civil War. The South supported itself entirely by taxes on cotton. They were all abolished again except capital [sic] taxes on whiskey, beer, and tobacco. A little later, to protect the farmers, these taxes were again levied on such things as oleomargarine, etc. In the Spanish War we reintroduced them on a large scale and during the last war to an immense extent. So that, you remember, at one time these internal revenue taxes yielded over a billion dollars. Now, of course, on account of the laws of 1919 and, especially, of 1921, these are very much fewer. The tobacco tax yielded about 176 millions of dollars last year; this figure also includes the state tax. Outside of tobacco the other revenues are insignificant. The vanishing taxes on automobiles yielded last year 66 millions of dollars, probably next year it will be nothing. The great difference between us and Europe is that we no longer get taxes from spirits. Our revenue last year amounted to 21 millions of dollars and permits amounted to $1,000, whereas before the war we would have three-quarters of a billion of dollars; this could cut our entire income tax into two. Tobacco is now our chief source of revenue. In the states, until very recently, we had no internal [sic] revenue. There have been two movements which are changing the situation. Just as the federal government is now encroaching upon the field of state revenues, so the states are encroaching upon what used to be considered the field of the federal government.

The one development is the so-called oil tax and coal tax in Pennsylvania. In Michigan the copper tax, but the revenue did not amount to very much. A more recent very significant development is the gasoline tax. Gasoline, from a revenue point of view, is taking the place of what was formerly done by whiskey. From the point of view of consumption in the wider sense, the gasoline tax is spreading very rapidly in every state in this country. The revenues are increasing enormously; last year they amounted to one hundred millions of dollars, and soon will be very much more than that. The reason is that the gasoline tax approaches more closely to the idea of benefit rather than ability, that is to say, it is becoming more and more felt that you can actually distinguish special benefit to the individual. You can do that on taxes in general but you have to have good roads, and the wear and tear of roads are due to the use of the roads and this is especially caused by heavier automobiles. Therefore the tax on gasoline is something entirely proper for those who use the roads, high powered cars paying their share. You also can put it in terms of the theory of ability, but only in the sense where contribution is a means of ability. This privilege of using the roads does, of course, enhance the ability of the individual. The supposition is that the man who owns expensive cars will be wealthy. That also explains this gasoline tax and even now we have a federal gasoline tax. That part of the ground that has been lost by the 18th Amendment is now being recovered by the introduction of this new system which does not meet [need in original] the same objection and works fairly well. If you add to this gasoline tax the other taxes that are paid by motor car users in this country today in some states, the revenue you get from licenses, you find that motor cars, entirely apart from other things, yield at least nearly three-quarters of a billion dollars. So the automobile industry is now replacing the whiskey industry as a means of providing revenue in this country.

Now the last point mentioned in this matter of excise taxation is the different methods that are employed. In other countries in the world there are five different ways. For instance, in the case of whiskey:

(1) Tax raw materials.
(2) Tax process. Even there you find different ways in different countries.
(3) To use the capacity tax, mesh tubs, etc. This method is practiced in Scotland.
(4) Tax the product, either for consumption or by putting a stamp on the barrel or by using a measuring apparatus.
(5) You get revenue by monopoly.

In the case of beer, you tax the raw material; tax the process of making the beer; you can tax the product by putting a stamp when it leaves the factory. Only in Germany do they do this. In the tax on tobacco you have three systems: (1) Either on the leaf, raw material, area of the ground, or weight, as the Germans do. (2) Tax on the product as we do. (3) Have a tobacco monopoly, as is done in a good many states.

For sugar, there are four methods: (1) Tax the raw material, sugar beet, or quantity of molasses. (2) The capacity tax, as they do in Belgium. (3) Density tax. (4) Taxing the product by using saccharimeter [saccarometer in original].

In the case of the salt tax, which some of the European countries still have, you can tax the product, as they do in a good many European states, or you can have a monopoly, as in Switzerland and other countries. You see, the whole tendency is to reduce the number of excises to the lowest practical limits and to use the simplest methods to get the largest revenue.

## 85. Customs Duties

These duties are chiefly not the export duties but, for a century or so, they are import duties. Before the war export duties had disappeared but during the war export duties were again introduced, such as coal in England, and some of them still survive. The import duties date from the Mercantilists; then they came in the 17th century from the idea of nationalism, protection, primarily on imports and, as you know, there have been various oscillations of the pendulum. England did away with protective duties, but since the war the system has been revived and you now have what amounts to a protective system in England, although only [not?] ostensibly. They are classified under three heads: (1) Given industries. (2) The system of taxes against dumping duties. (3) Duties have been imposed in order to correct for certain changes. England now having lost the monopoly of the world's market and having thus changed from a country where free trade was inevitable and profitable, she is now [not in the original] going back to some degree to what you find in other countries.

Customs duties originally came from export duties, then from import duties. The reaction in the 17th century and, especially, the 18th century was such that the import duties under the sway of protective considerations reached to very high figures. The reaction began in England about the middle of the century. The significance of the subject today is that import duties have varied in weight according to the prevalence either of the revenue question or protective questions. Wherever the revenue questions are important the number of import duties are very much reduced. {Whereas in a great many other countries they follow this, the reverse is true in this country.} The problem is primarily economic, and as such it transcends these particular lectures. For the chief revenues that involve fiscal considerations, the question is, whether we should have ad valorem or specific duties.

We will limit our consideration to a few words on the question of ad valorem and specific duties. One objection to specific duties is that they are unequal, that they are heavier upon cheaper commodities. The second objection is that duties remain the same even though prices change from month to month or from year to year. The third objection is that specific duties are very difficult to fix intelligently. The fourth is that their incidence is apt to be concealed. The objections to the ad valorem duties, on the other hand, are: First and foremost, that they lead to under-valuation and fraud. Second, that it is exceedingly difficult to administer the duties intelligently, because every official has to be an expert on either all the tariff laws or particular commodities. Of course, we haven't such experts. The consequence is that both systems have advantages and disadvantages. Specific duties are always protected by the advocates of the tariff and ad valorem duties are protected by free traders. At first, all the duties are ad valorem, then you have far more specific duties. If, however, apart from political considerations, you should attempt to lay down principles, then the considerations would be the following: (1) Is the article homogeneous or not? Take, for instance, pig iron and compare it with silk. It is very easy to tell about pig iron but duties upon silk are so difficult to determine, you have so many varieties. (2) Are the duties collected along a very long land front or in a few seaports? That is a very important consideration. (3) Most important, what are the characteristics of the administrative system? The better your administration, the greater chance you have to succeed with an ad valorem system. With poor administration, it is more difficult.

We generally compromise. Our tariff is filled with mixed duties. In the last few decades, we have all the specific duties. The tariff at one time played the sole role in our federal system. You remember that from the end of the war with England up to the Civil War we relied exclusively upon the tariff, then afterward we relied chiefly upon the tariff more for internal revenue. For the last decade or two the import duties have been of very slight significance. The tariff in the future will be of very slight importance from a fiscal point of view, and it is becoming continuously less important even from a general economic point of view. Customs duties will become of comparatively small importance relative to the other sources of revenue.

## 86. **Stamp Taxes**

## 87. **Taxes on Transactions**

## 88. **Sales Tax**

Stamp taxes, of course, no longer mean what they originally meant. Stamp taxes, until very recently, meant taxes on transactions, but you remember I told you in the last lecture or two that some of our excises are now also paid in this country by stamp, such as the whiskey tax in the past. The tobacco tax is still paid by stamp, and you would not call the tobacco tax a stamp tax.

This method was brought from Holland to England, which developed the system of stamp taxes and has kept it ever since. The chief characteristic of the English system is the so-called penny stamp [tax in the original] as an everyday receipt. Of course, they have all kinds, which bring in a quite substantial revenue. In France, the stamp tax has been of much greater importance, especially under the name of registry taxes, which have to be paid by means of stamps. This is one of the four great sources of revenue. In other countries you also find stamp taxes quite common.

In this country we used it in the 18th century when Hamilton required revenue. We used it during the war with England, the war with Spain, the Civil War, and, especially, the recent war. They are of two schedules: Schedule A, discriminatory, and Schedule B, on cosmetics, taxes on sales. Those discriminatory stamps have brought in very large revenues. We have had at other times a few such taxes. One of the chief source of controversy in this country was over the check of stamp taxes. The argument there is different in different sections of the country, according to where and how this is used for daily payment, etc. It was, however, abandoned. The revenue from stamps, however, has always been of minor importance, and as soon as the exigency disappeared these taxes also disappeared. The states where you have these taxes are Pennsylvania, Virginia and Maryland. One particular form of stamp tax we have in New York and Massachusetts is the tax on the exchange of stock. As to how far taxes ought to be extended to the transfer of bonds as well as the transfer of stocks, and whether taxes should be imposed upon normal [par] value or actual value, we haven't got time to go into detail. I will say, however, that this stock tax brings New York such a large income that it will no doubt remain. Outside of this you have very few stamp taxes. We had a very great political controversy as to whether this tax should or should not be extended under the name of a sales tax. The sales tax is found in countries only when they reach their end and only when every other means is exhausted. This was especially true in Germany. Other countries, which are little better, turn this down, such as England and Italy. With us it came [up] when the income tax was being introduced. The income tax looked so high that anything looked better for us then. It was opposed not alone by the Secretary of the Treasury but by most scholars, and never had any political chance at all. They asked me to come down and address [the issue]. There was great excitement. It looked at one time that a sales tax would go through, but all arguments were clearly opposed to the sales tax. It was shown that this was entirely incompatible. They had it in France and Germany, but they gave it up at the very first opportunity.

## 89. **Taxes on Transportation and Communication**

They are the marks of the primitive economic stages. When the African explorers went to explore parts [ports in the original] of Africa, whenever they entered into the domain they had to buy their way through. We find this all through the Middle Ages – interior duties, internal taxes. They were very common on land and on water. On land those taxes were at first on human beings, but they gave way to taxes on transportation of vehicles. In England, the post office system developed, Pitt introduced taxes on coaches, but when these coaches ran on iron rails they were then transferred to railroad taxes. So, in England, we find taxes on railroad passenger tariffs [passage duties in the original]. Of course, in times of distress these taxes are always increased. On the continent, you find taxes also on freight. With us, the chief examples of taxes on transportation are seen in the Civil War, when we put taxes on pretty much all media of transportation – railroads, steamboats, telegraph, express receipts, tonnage, etc. During the recent war these taxes were very much increased and brought in quite a number of millions of dollars, but they were the first taxes to go. With us now, as well as pretty much all other advanced countries, taxes on transportation and media of communication are reserved only for very extraordinary exigencies, so that they will not be of any importance under normal conditions in the future. You might put under the same category the taxes on motor transportation: gasoline taxes and taxes on motor cars. They are under the excise rather than transportation.

So that casting a glance over the system of indirect taxes, they are limited to a few widely important excise taxes and to import duties. In the Romanic countries, you both have a much wider system of these taxes on transactions and consumption and find that direct taxes are slowly finding their way.

## Part 3. Local Taxation

### 90. **Separation of Sources, Division of Yield, or Apportionment** [**English System** in the original]

In the last few years very much more attention has been paid to this subject. The problem has come up in federal countries and in unitary countries in connection with local taxation. Local taxation is becoming more important, especially as aftermath of war. The problem therefore is being studied afresh. A great commission in Australia studied this problem, another commission in India studied and brought volumes of reports. The same thing took place in South Africa. Without going into detail, we might say that there are really three important systems [questions in the original] to consider.

One is the system of separation or segregation of the source of revenue. That was the system in this country. For instance, all indirect taxes were taken charge of by the Federal Government and all direct taxes by state governments. When it came to the question of the relation between state and federal governments, this was a different question. The doctrine of direct taxes was very popular a generation ago. A generation ago the general property tax was universal and all attempts to attack the problem frontally seemed to be impossible. When it came to consider the situation in New York, the only way out of the difficulty seemed to be to develop the ideas of separation, whereby the states would no longer rely upon the general property tax but satisfy themselves with corporation taxes and the inheritance tax, and that the general property tax should be reserved for localities. You know the reason why. That was the only way out. It enabled us to take a great step forward in New York State, then California took the matter up and carried out the system of separation much more fully. Although it marked a decided step forward at the time, it is no longer either needed or advisable today, because it is being replaced by the income tax and under the income tax you don't need this [a system of separation]. Professor Haig [Hays in original] was called again to help to form a new system. If you rely on the tax on the corporation and on the inheritance tax, with the growing needs of local governments you will got into difficulty, as this will involve imposing higher taxes on corporations. We will see what the outcome is going to be. In some of our middle and Southern states, there is still a great movement in favor of separation.

The second plan is what might be called the division of yield. To have one form of government, state or federal, all over collect the taxes and then divide it up with each other. I will explain what I mean. In Germany it became necessary to centralize their administration and all taxes are now collected by the central government alone. Whereas in Germany before this, it was the other way. Taxes were collected by states and then certain parts were handed over to the federal government. In this country we had that same system. In 1789 all the revenues were collected by the states and handed over under the name of requisitions to the [federal] government. Now in our state-local taxation in most of our states we are the other way around and taxes are collected entirely by local government, and then certain proportions go to the state government. Now, both of those systems, of course, are bad. The German system is better but would not do for this country. And the other side, collection of all taxes by the local government is, of course, breaking down. We have developed in this country for the last ten or twenty years a method considerably approaching the German system, or the Canadian system.

The only way for England to get control of the situation will be by means of these grants that she gives and exercise general control. We find that this system has been most popular in the South. But human beings are the same all over. With the growth of national economic interests over against state and local interests, you are going to have more expenditure and ultimately more control.

In many ways we are therefore approaching in some respect the system which is found in Canada, to a certain extent in Australia and in other federal governments as well.

The third system is the one which, on the whole, seems to be the best suited to the modern condition, and that is neither separation of sources nor the division of the yield, but apportionment. Explaining what I mean: In our federal inheritance tax, money is collected by the federal government yet 80% of it goes, according to the law, to the state. The same thing is true with our income tax in this state. We have other taxes whereby tax revenues are apportioned by the law and certain parts go to states. That means that you no longer would have complete separation of sources and also breaks down the argument commonly heard today that the federal government is encroaching upon the domain of states. There is nothing that naturally belongs to federal or state government. The whole field is open. What is going to happen is that a further study will be made of the local forces, federal forces and state forces. You cannot say what is going to be the best. It depends on the kind of tax, and great study will be made in the future. For instance, how best are you going to reach the ability to pay of railroads? If you take the earnings, only the federal government can reach this. That problem, therefore, is one that remains to be solved in the future. Lloyd George[204] was just getting ready to attack that problem. He was working on a scheme for a division of the income tax in England. The net result is that you cannot have hard and fast grants as regards local, state and federal finance; the problem has to be solved in each case, then abandon either separation or division.

# PART III

# PUBLIC EXPENDITURE

## 97. **General Considerations**

Now, as the time is getting short,[205] we will omit the statistics and limit ourselves to the consideration of the theories and general conclusions of the subject.

If you enter into the details, you get into problems concerning the expenditure of every item and then you will have no distinction between administrative science and fiscal science. At the same time, there are certain points of scientific importance which belong to scientific finance. It is from that point of view that I shall discuss them.

Now one of the chief problems in this point of view will be the economic effects of public expenditure. Here you find a gap in the books, and very largely for the same reason you find a gap in consumption. You see, public expenditure is public consumption; therefore consumption in economic theory ought to be divided into private consumption and public consumption. Now, if you do that you find even in private consumption there has been very little distinction. In my book on installment,[206] I find I had to develop an entirely new theory and it seems to me that the analysis there is applicable to this matter. You remember the distinction between productive and unproductive consumption, and that I pointed out that there are four different phases of production or of consumption. If you sum both up, the utilization of wealth, there are four different ways you can utilize wealth, e.g. positive, neutral, wasteful and destructive utilities. Positive being the bringing up the body of a boy. Neutral can be applied to that of man; wasteful consumption, when you eat more than you ought to from the point of view of unnecessary wasteful consumption; and destructive consumption being indulgence in poison, such as alcohol, destroying the tissue. Those are the points of these four aspects that I think are very fruitful in dealing with public finance and it applies in entirely the same way. {How often you read in the papers to be economical for building roads, expenditures involved that they are bad things, etc.} Now if you apply these four tests you get your answer. If the money is spent for wasteful and destructive purposes, then, of course, it is a bad thing, and if public expenditures necessarily involve waste, where private consumption would involve no waste, then comes the argument against government dealings which becomes perfectly sound. You must remember, however, that public consumption, like private, may be positive or neutral. If it is positive, there is no objection. If the government is going to build schools and good roads, or if it does things which perhaps cannot be done so well by private individuals, as for instance, the building of the Panama Canal, which could not be built when tried by private companies, and government takes charge, this is an example of positive consumption and money spent in this way is far from being unproductive. You see, this applies to public as well as private expenditure. The whole problem of government expenditure can be approached from this point of view, which I think is not only new but a perfectly legitimate defensible point of view. From these you can reach to Adam Smith and a great many others. I cannot dwell upon that further. As to the expenditure, of course, it depends upon financial needs. When the financial needs arise you have (1) economic and social conditions, (2) political conditions, and (3) conditions with the forms of government and questions of centralization and decentralization. All these we have adequately touched upon in the beginning of the course.

## 98. **History**

About the history of the growth of public expenditure, without giving you any facts, I will just consider the main causes of the increase of public expenditure: Nominal increase: The amount of money that is spent is, of course, very much greater than before the war, but that does not represent the real expenditure, because there has been a change in the standard of value. Many people forget that. Of course, expenditure may change without having a change in real expenditure. Second, real expenditure: You find there two very decided movements, the proportional and the progressive. That is to say, expenditure increases proportionally to certain things. What are those things? They increase, first, with the growth of population; second, with the growth of territory; and third and most important of all, with the growth of wealth. The ordinary figures you can find in all the books. Per capita expenses have very little use. In addition to these proportionate increases, you have those expenses which grow progressively larger, relatively larger, which take a continuously greater portion of social income. Where income grows, expenses grow with it. There are also other things, such as governmental inefficiency and waste. A far more important thing, however, is the influence of war. Feudal methods of making war before the development of gunpowder required less expense than modern warfare when factory production is applied to war. Of course, you will spend more and more of your income. Another point is the expansion of the functions of the government. That is due to the growth of democracy, also due to the growth of capitalism. Also, you remember, the gradual change from repressive to preventive, from preventive to ameliorative, and from ameliorative to constructive stage.

If you look at it from another point of view, you may say that while there is always this tendency for an increase of expenditure, there is always action and reaction in life, and that whenever you see undue increase, you find an attempt to put limits upon expenditure. This is the reverse movement, and the attempt again has three aspects, which tend to keep expenditures down. (1) Movement designed to cut out waste and inefficiency. This explains the budgetary movement in modern times during the last decade or two. (2) To lessen war, therefore diminish expense – efforts at prohibition of war are very perceptible at the present time. (3) Most important of all is the tendency to carefully scrutinize the objects of expenditures and to introduce the very salutary method of constant comparison between private and public expenditure. All in all, in spite of the different influences, there is nevertheless a continuous increase in public expenditures, both absolute and relative, taking place at the present time. In the future still larger importance will be attached to public consumption than private consumption.

## 99. **Principles**

What can be said about principles? First, of course, we would say expenditures must be productive, positive, or creative. And, again, you come to the analysis that I made in that book on the installment plan. You must accept the three-fold analysis of production. Second, the peculiar concept of wealth, productive when money is made from the enterprise. Third, an more important conception: Each of the first two may be resolved in the modern analysis of economic factors in which you conceive of wealth as simply affording satisfaction. Translated into ordinary terms, it means that the expenditure of government must be gratifying, positive, and also bring about immaterial results as well as mere material results. This seems to involve a great many things. Certain things may seem destructive, yet really be constructive in this other point of view, such as when you destroy and consume coal in a locomotive boiler you are creating something more worthwhile than destroying the coal.

The second principle is that expenditure must be for the general benefit. By that I don't mean general in the sense opposed to local, nor do I mean general as opposed to particular. The expenditure may involve profits to a particular individual or classes. What I mean by this is, of course, what constitutes a legal answer, that there must be a public problem. The third point is that the expenditure must be necessary. Without going into detail, I think we can sum it up in a few words: government is always entitled to do what private individuals don't want to do, or that they are not able to do, or what they ought not to do. An example of what they don't want to do is private subways, etc. Second, what they are not able to do, such as the building of the Panama Canal. What they ought not to do, deals with the matter of justice. In general the burden of proof is always {after the community upon the public.} In Russia, the argument is the other way. They say that the private individual can do only what the government cannot or does not want to do; our civilization is the other way around. Government will do these things because of the three reasons.

The fourth consideration is that the expenditures must be economical, but that does not mean niggardly. There is a difference between economical and niggardliness, and democracies are usually niggardly. The diplomatic service, for instance, in democratic countries is paid absurdly small. The fifth question is, that the expenditure {must balance definite proportion to social income.} Extravagance is absurd whether in the case of private or government [expenditure].

## 100. **Classification**

How are you going to classify the expenditure? There are two ways in which you can do it. The most scientific way would be to classify it according to the aims and objects of the government action. Here we have the four answers that we pointed out. According to the actual forms of government, and you have to divide them into five criteria. We do not have today any general widely accepted classification applying to the whole country. Countries differ so widely among each other, and yet classification is the very first step in scientific process. I shall, therefore, suggest, as summing up comprehensively the following five classifications:

(1) Expenditures for maintenance. These would be the general administrative expenditures.
(2) Expenditures for protection. These would be divided into (a) defence, and (b) law and order. The first includes the army and navy; the second, the judiciary, the police, the general regulatory expenses.
(3) Improvement expenses. There can be divided in (a) education, including science and art; (b) transportation, including roads; (c) charity and health, including sewer and recreation; and (d) insurance and pensions and things of that kind.
(4) Productive or creative expenditures. Creative, really productive of wealth are the economic expenditures of government, which are either (a) indirect, through bonds and subsidies to improve production; or (b) direct, where the government takes care of public enterprise, spends money, and very often has deficit, {designs so.}
(5) Public debt expenditures. A large part of these go to debt for the problems of public improvements, such as roads and education.

Now if you make this classification, you will find that every kind of expenditure can be put under one of these divisions. Taking them altogether, however, and arranging them in the order of fiscal importance, you have to place them in the following way: Most of our expenditures are: (1) productive; (2) public debt expenditure, (3) maintenance expenditure; (4) (unfortunately) improvement expenditure: (5) least of all, in connection with economic expenditures.

## 101. A. **General Governmental Expenditures**

These expenditures are either executive or legislative. In modern times the executive expenditures are of very minor importance. Of course, in former times, under monarchy, they were of great importance. The King of Siam receives 312 millions; the king of Italy, $1\frac{1}{2}$ millions; the King of England, so much. Now those figures are very high compared with the salaries of presidents. It was only a few years ago that the salary of the president was increased. The salaries of public [the word "utility" was typed and then x-ed out] presidents are far less than the salaries of industrial presidents. When you come to legislative expenses, there again it depends upon your form of government. In some countries, legislators are not paid anything.

## 102. B. **Protective Expenditures**

Two questions are involved in this: one is the protection of the community and the other is the protection of the individual. The protection of the individual is accomplished by the police, and, so far as property is concerned, by the fire department. I will say a few words about general defence. The expenditures for defence were originally the expenditures of war. But more and more these have become outweighed in importance by the second class of expenditure, that is, preparation for war. Therefore, we have war expenses and preparation-for-war expenses. Here is a great change due partly to the increase in carrying the warfare and partly also to the change in the methods of warfare. Historically, in Adam Smith you will find that we have gone through a rather interesting development. Originally war meant *levée en masse*.[207] You see this in the early stages in Rome, etc. Then you get into the stage of the division of labor, where only a part of the community devotes the energy to such ends, and you have militia, or you have paid mercenaries. Then you have the gradual development again where, as in the original stage, you have conscription of the whole working male force, and finally, as was the case during the last war, the war was between peoples, the whole nations took part. So the technical changes of warfare bring about expenditure which is relatively enormously larger than those of the Middle Ages. In these [possible word indecipherable] days expenditures during the war period run to millions of dollars per day. The ostensible prosperity during the war does not enable the war to go on very long. But what remains true, in the economic view, is that if modern wars last a long time, they will ruin the civilization. The expenses of the last war are responsible for the whole burden of today. What money we made during the war we spent during the war. The movement which is now going on all over the world to prevent war, is perfectly explicable from an economic point of view.

The other side of it is more important, and that is the expense for preparation for war. It is thought that best way to prevent war is to be ready for war; therefore, we spend more and more money. So that a large part of government expenditure goes to our army and navy. That is again wrong. Why are these immense efforts made by all governments? As a matter of fact, we spent more money than any other country; relative to our wealth and population we spent less, but absolutely we spent more. Of course, this is due to the fact that we pay large monthly salaries to our Army, whereas in Europe they pay less.

Moreover, the matter of pensions also must be discussed here. There we come to a subject on which, under ordinary conditions, we did not spend much; but as years went on, we spent more and more. This was due to the widening of pensions. At one time we spent 175 millions of dollars. The new system during the late war was a very much better one, even there you remember the expenditures ran up to several hundred millions. The total amount of money that we spend today is about half a billion dollars. This expenditure still exists in Germany even though its standing army has disappeared.

The place that the army occupied is now taken by the second class of protection, police; and while the expenses for general protection were greater than it used to be, police expenditures are something entirely new. This is not yet a century old, inasmuch as the idea of police came from England. Robert Peel[208] first devised this plan of protection in cities to individuals, which is why in England they are called Peelers [bobbies]. The police have developed from this very modest beginning to the very important position they occupy today, whereby we in our cities have expenditures for police greater than anything else, except education. The need for police vastly multiplies, and it has developed from a local function to a state function; and in great many countries of Europe the army is partly utilized for police protection. Sometimes we use our militia in order to protect the mail and also during strikes. You have to be very sure you understand well the limitations which differ from country to country as to the relative expenditures for the general community and the individual.

In recent years we have had another great extension of police in this country. We are using the word in its original sense, the sense that Aristotle used and the sense enacted in the Holy Roman Empire, namely, police ordnances that, in this country, go under the name of the police power, which means administrative power. If you include all these different phases, you find that the expenses connected with our protection system are rapidly increasing. You might even put under this head expenses connected with our commissions: bank commissions and public service commissions and other divisions to protect the interest of public, etc.

## 103. C. **Judicial and Reformatory Expenditures**

Originally justice was very largely the result of voluntary arbitration, but when the government or king took the matter over it was an expenditure. The king sold justice, and for that reason it was made compulsory. After a long time, recognition of the social function of justice appears and expenditure increases. It is only in the very late period that judges are paid. Adam Smith, only 150 years ago, says that judges have to get paid money so that they will work harder. Only in the last few years have they gone to salary. Salaries paid to judges are exceedingly small, as, for instance, to the members of the Supreme Court. You can never get judges for the salary, except for the honor. We have a long way to go to get to the English position where their judges are paid very high salaries. The result, therefore, is that expenses for justice are increasing rapidly and will continue to increase. Without going into detail, it may be said that the expenses of justice are bound to grow with the progress of society.

What is true of justice is far more true of expenditures for prisoners: reformatory expenses. That is due to a complete revolution in our attitude toward the whole subject of punishment. The idea is that through proper treatment something may be done to remedy the situation of an individual who is often considered to be a criminal. All these ideas simply mean increased expenditure. This idea goes hand in hand with the relatively new forms of charity organization. Hospitals did not exist in ancient times; even among the Jews they did not have the idea of a hospital. This is a Christian idea. It was a function of the church, and it is only with the separation of church from state that modern expenditures for hospitals have been increased. Then comes the other side of the question, biological observation: with a more complex social system we will have greater number of diseases. Of course, science is getting rid of certain diseases. We no longer need to die from yellow fever, etc., but for every disease we cure, new ones appear; the proportion of invalids is greater today than in ancient times, due to heart disease, insanity, etc. With the recognition of that fact it is perfectly clear that our expenditures are growing. The immense expenditure in New York State is largely ascribed to that fact.

## 104. D. **Educational and Charitable Expenditures**

It is only with education that we finally hope to get rid of these evils. These have gone through the same fiscal stages that I mentioned. At first, it was private expenditure; schools were paid for by the individual, there was no public expenditure. When government took over the matter of education, there was a new expenditure. In this country we find three stages. In the East, most of our educational institutions are on the whole private institutions. Private schools are, on the whole, the best schools. Universities are best when they are private, like this one and others. The further west you go there is more of a feeling of democracy and the number of public educational institutions is larger. You have the very same system in Switzerland and in Germany. Then, finally, in the last stage, you notice a very great change in the public high schools throughout the country and in our state universities. The public schools were due to the Dutch, not the English; the system was brought over from Holland. We are now beginning with free schools and free textbooks; millions of dollars are involved in this. In some states, we have printing plants and government-paid writers. In some countries [counties?] they even give free breakfasts: in other places, they even give them lunch; and in some places the socialists go so far as giving private clothing. All those developments of educational expenditures depend not so much upon insular features as they do upon the widening scope of education. We are expanding the scope of education. Then, recently you have the idea that it is absurd to limit a person to certain years of education, to 20, 30 years; the whole society recognizes the fact that education never finishes, you can never learn too much.

## 105. E. **Health and Recreation Expenditures**

By health expenditure we mean expenditure for both conservation of health and then, of course, in addition, sanitation; this, insofar as municipal expenses are concerned, includes the whole subject of street cleaning and sewers and sewage disposal. These expenses also are very large. With reference to recreation, we have to include important topics: the origin of parks, gardens, bathing beaches, baths, both summer and now also winter, and the development in recent years of certain celebrations and pageants. Altogether, these expenses for recreation are comparatively small, although bound to grow in the course of time.

## 106. F. **Commerce and Industry Expenditures**

In some ways the most important public expenditures are the expenses connected with commerce and industry; that is to say, expenditures dealing with economic production. These are of two kinds: indirect protection, such as bounties and subsidies; and direct expenditures, including enterprise. The first category, bounties and subsidies, has a very long history. Bounties are really of four kinds: (a) Military bounties, such as we had in the Civil War. (b) Forest bounties, which we find very widely developed in this country. (c) Agricultural bounties, which have a longer and more interesting history. During colonial times, these bounties were given to all kinds of agricultural products, such as flax, hemp etc. In modern times, the chief ones are to sugar, also silk. It dates back to Napoleonic times. International conferences were held and sugar bounties were finally abolished a decade or two ago in this country. The chief example, however, of bounties is (d) in connection with transportation, either for land or for shipping. Land transportation bounties have a great history in this country, first for roads and canals, and then railroads. Many of our railroads, like the Erie Railroad, were constructed with direct money bounties. Others, such as the Pacific Railroad, were granted land and land-related privileges. That is now all a matter of the past. Shipping subsidies are of different kinds: simply for so much money, others like the mail subsidies that England gives, and subsidies for construction. We have had subsidies of every possible kind for half a century, as Japan has also had at various times since the end of the last century. The problem has come up again in this country because subsidies are not very successful. In connection with the disposition of our national Merchant Marine, owing to the new laws, it must be absolutely impossible for operators to compete with vessels of other countries where the standard of life is so much lower. The problem involved here, however, is really not fiscal; it involves the general question of protection. Those who believe in the doctrine of free trade oppose it. Under existing conditions, government has to either manage it itself or spend, probably, a small amount of money. If we want a Merchant Marine, these two forms of expenditures are inevitable. Today, then, the question of bounties and subsidies is comparatively slight.

On the other hand, the greatest expenditures is for public enterprise. We discussed some of the phases of public enterprise under the heading of revenue, because you get a revenue from some public enterprise. In most cases, however, these public enterprises call for expenditure rather than revenue. Take, for instance, what happened with the expenses of transportation: The high roads that have come everywhere in these stages. First, roads were in private hands in Asia. A good many of the noble families started out this way. After the first period, it was used for the purpose of extortion; some of our private docks are in this same situation. Railroads are also regulated this way. In the second period, they became semi-public, such as the period of highway tolls in England in the case of turnpikes. These tolls continue, to keep the roads in repair – originally the result of unbearable conditions throughout the Middle Ages. The third period is when the government takes them all and uses them as a means of profit. Well, that, of course, was true in a great many countries; it was true as regards railroads in Prussia. The fourth stage is when no profits are made, but the cost is covered by the principle of fees, with tolls. This is the system in the Panama Canal. The fifth stage is when cost is only partially covered and expenditures [beyond toll revenues] are needed; that is true in some of the bridges in this city. Finally, comes the last stage, with no charge at all, like the Erie Canal, and as is true about ordinary streets. There are a few places in the South where they still charge tolls. Whether this is a fiscal problem is a very great question. There has been this historical development, however, from the first stage of private extortion to the last stage of large public expenditures. This differs from country to country.

The other categories of such expenditures deal with the following:

(1) Transportation.
(2) Expenditures connected with trade, such as coinage, and weights and measures.
(3) Expenditures connected with natural resources: (a) Irrigation, drainage, nominally as we only have slight net expenditure. (b) Forestry, there is still some form of expenditure but some day will bring in revenue. (c) Connected with rivers and harbors – Mississippi scheme. In the case of harbors, the expenditures are to be met by fees and tolls. In the case of highways, however, we have very large expenditures running into billions of dollars. You remember why our railroads cost so little: because they were built with[out] grade crossing, but everywhere in Europe this was not done. Now we are doing this and it is costing hundreds of millions of dollars. It is very easy to picture that the situation of a total capitalization of 20 billions of dollars would not be less than doubled before we get to this point. (d) Connected with industry. We discussed the revenue side of it under the heading of government monopolies, but in a great many cases we have them for other reasons. Take our shipping board: its cost goes up to hundreds of millions already, and expenses of that kind make it rather unlikely that we shall continue to have great expenditure for exchange purposes in the future. The whole tendency is to turn this over to private management because there are no good adequate social reasons for

government management. The situation is now very different from what it was a few decades and a generation ago. At that time, it was private management versus government management; now it is government management, by government or social control. In a great many places we find that adequate results have been achieved by social control, where the situation is complicated. The chances are that we shall before long have a government system and then the choice will have to be between paying tolls and preservation for social reasons and for [with a?] deficit? That is a political problem. So far as the economic problem is concerned, where you have expenditure everywhere running to hundreds and [of?] millions of dollars, the present tendency, and I think it is perfectly legitimate, is to put part of the burden upon those who are using the facilities, which means, therefore, the argument will be a strong one for an increase of fares in the matter of subways, etc. Where, however, the argument is social, the political argument seems to oppose that. Then, of course, we have to have a very great increase in government expenditures for this purpose. In one way or another, the whole category of expenditures may be increased in the future, relatively.

## 107. G. **Public Debt Expenditures**

## 108. **Conclusion**

Then we come to the conclusion, which is that perhaps the first problem connected with expenditures is the need for reform in our attitude toward expenditure: the distinction between productive and unproductive expenditure, recognition that certain expenditures are needed and will be needed more in the future, and that the {tax limitation laws are considered}. The second problem is that of centralizing the responsibility for expenditure instead of the present-day diffusion of responsibility. That brings us to the whole question of the budget. Then again, perhaps the most acute problem in this country is the question as to how are you going to combine [revenue sources]? Are you going to pay from your annual revenues or pay from borrowing? The difference of expenditures, it must be said, lies in different political constitutional conditions and especially in economic conditions.

# PART IV

# PUBLIC CREDIT

## 109. **General Considerations**

When dealing with public credit, we ask, what is credit? We have no time to go into economic theories. We can define credit as a phenomenon of all transactions: something is done now for something else in the future: a transaction the completion of which is deferred to the future. The interval brings the question of trust. There again credit is being utilized for the purpose of: 1. To cover temporary shortage. 2. To meet a sudden emergency or outlay – here you have to give security. 3. Increase the productivity of your business. The whole 19th century development of banks and bank credit. 4. The most recent development of credit as applied to consumption, installment and things of that kind. Government credit, or, of course, public credit, is like private credit in all respects except that it is the state and not the individual who is asking for credit, and the only differences, which are very important, between private and public credit are as follows:

(1) The state is sovereign and therefore can repudiate these debts economically. It is not correct to assume that government cannot borrow any more when it repudiates. Many public debts have been completely repudiated in this country in the 30s and 40s.
(2) Government has eternal life, whereas the individual dies; therefore government has a much longer period to pay the debt, and also there is the question as to whether the debts have to be paid at all. Many European countries don't pay their debt at all. Also railroads never pay their debts.[209]
(3) Government has no property for mortgage collateral when it borrows money. In ordinary times, the government has very little property to mortgage and the only thing it could mortgage would be the annual revenues. But, of course, the difference between government and the private individual is that government can increase its revenues easily, the private individual with much more difficulty.
(4) Public affairs are open to inspection, whereas private affairs are not.

## 110. **History of Public Credit**

The real beginnings are found in the Italian and Spanish domains early in the Middle Ages. Public debts were the private debts of kings. That was true in England in the 17th century. The bankers made all their money, sometimes lost all their money, by lending to kings who paid when they liked. The idea of public credit over against the kings' credit developed in Italian towns, such as in Geneva and Naples in the 15th and 14th centuries. During wartime they needed money, and they singled out the first hundred wealthy citizens. They compelled these citizens to loan money, in return for which they received shares in associations which were formed by these citizens. Repayment for these was so arranged that every year a certain percentage would be paid together with the interest and in the course of time the whole thing would be paid. The way this association got its money was that certain taxes were turned over to its disposal.

In this origin of public debts you have the beginnings of three very distinct institutions:

(1) You have the corporation, because in the association every man had a share – the first form of corporation, with liabilities limited.
(2) Beginnings of banks, because these corporations were given the name bank – bank of mud or sand bank.
(3) Public Loan. The annuities were called *loca montie*.[210] There were different kinds of annuities, running from a short number of years to life annuities.

The great problem arose in the Middle Ages, whether this interfered with the usury laws. And this entered into scientific discussions. National conferences took place. But finally the Popes themselves needed money and they began to borrow, so that distinctions were made.

That was the origin of public debts, and you see it had its origin primarily in war, the debts were war debts. That still is very important; during the last war we spent about 240 odd billions of dollars. Apart from war, the great increase of modern debt is due to other things, such as debts incurring in productive enterprises, for instance, the Panama Canal, building roads and schools, or the Mississippi scheme now.

## 111. War Chests and Reserves

Although a discussion of this is scarcely needed now, the following was the case in the past: What can you do if you don't want any debt for war expenses? There is only one alternative because, as we shall see in a moment, the attempt to be able to pay for war is familiar. In ancient times, they had these war chests. In time of peace, every government heaped up a mass of metal, silver or gold, to be utilized in time of war. As in the cases of Greece and Rome, they had special places to keep this. Throughout the Middle ages that was the custom. In modern times, the chief survival of that was in Germany, where in Potsdam a large part of the money that they received as an indemnity from France was kept, several millions. Even Professor Wagner, in his *Science of Finance*,[211] seems to make a strong argument for such a measure. This idea was abolished, however, as a result of the last war. In the first place, in order to have a fund which would be adequate, you must have an immense sum. Therefore, you will be simply wasting your energy by gathering an immense mass of bullion. You can get the same result by giving bullion certificates. The only thing you must do in time of emergency is to tax the people adequately or borrow, and therefore public debts become ultimately important.

## 112. **Theories of Public Credit**

We will say a few words about just a few facts, although most of these you can get in the books. Speaking of the United States, our first public debt amounted to $75,000,000. It was gradually paid off; in 1812 it was reduced to $45,000,000. As a result of the war with England, the debt increased to $127,000,000. That was gradually paid off, and in 1835 we were practically out of debt. Then came the bad years, and we had to borrow money to keep going. In 1846, we owed $15,000,000. As a result of the war with Mexico, we owed $68,000,000, which was gradually reduced until the beginning of 1857, when we owed $28,000,000. Then, owing to our bad policy, we had to borrow money. Before the Civil War we owed $50,000,000; as a result of the Civil War, we owed almost $1,844,000[,000]. The low water mark was reached in 1893 when we owed $1,500,000,000, but of that sum $585,000,000 was due to interest. Then came the war with Spain, and as a result we reached the high point of a little over $1,000,000,000 [?] total interest bearing debt. It remained for several decades at that for {some bonds that banks could invest}. In 1918 it was still $971,000,000. Then came the rapid increase and the high water mark was reached in 1919; it amounted to $25,250,000[,000]. Since then we have been rapidly paying the interest bearing debt, which is now about $18,500,000,000. So that is the history of the federal debt in this country.[212]

As regards other countries, debts before the war, in 1913–1914, were as follows: France, about $6,250,000,000; Russia, $4,500,000,000; Great Britain, under $3,500,000,000; Germany, about $1,250,000,000 for the federal debt and $4,000,000,000 for the states; Italy, a little under $3,000,000,000; Spain, a little under $2,000,000,000; Japan, $1,250,000,000; India, about $1,500,000,000. As a result of the great war, of course, the German debt was wiped out; everybody who had invested lost. The debts in Great Britain ran up to $37.25 billion dollars; France, 32.25 billion dollars, somewhat more now; Russia, 25.5 billion dollars; Italy, 13 billion dollars. You can count the German debt, before it was wiped out, amounting to 48.5 billion dollars. Japan has a very small debt. Debt in other countries heaped up as a result of the war amounted to 140 billion dollars. So that is the amount of debt weighing upon the world. That, of course, is a very important problem.

Now we come to the theory: It was not until the end of the 17th and the beginning of the 18th century that the discussion began, but the real discussion began at end of the 18th and the beginning of the 19th centuries. At the beginning, everybody looked upon public debt as creating new wealth, just as private debt did. (You remember what Law[213] did in France.) The high water mark of this tendency was reached in the book of Pinto[214] in his great book on credit, praising credit to the sky: "National debt, national blessing". Melon said: "National debt never imposes any burden, because it is a debt from the right hand to the left hand".

The second view marks the reaction, and this took place owing to the great increase of debt. This tendency reached its fullest expression at the time of David Hume, who gave five or six reasons why public debts were bad. He influenced Adam Smith so much, he opposed public debt.

The end of the century marks the third movement, when we get the idea of a sinking fund. Sinclair[215] and Price[216] rather favored debts. The 19th century discussion began with Ricardo in his discussion of the financing of the Napoleonic war.

The fourth movement was reached in the middle of the century by such writers as John Stuart Mill. He was very much opposed to public debt, in connection with working people's attitude toward the wages fund doctrine.

The fifth movement came as a reaction, especially in Germany. Two writers, Dietzel [Ditzel in original] and Nebenius[217] [Nebetian in original], favored borrowing not only in time of war but in time of peace, and carried the doctrine to a rather absurd extreme. So that we now have come, especially as a result of the last war, to be able to give a correct appraisal of the subject. We realize that the real theory of public credit as a public function lies in these two facts: (1) The doctrine of marginal disposability: When you invite subscription from people who lend money to the government, it leaves in the hands of those who can make better use of the funds the capital which otherwise would have to be taken by taxes. By taxation you are imposing much more burden, whereas if you invite voluntary subscription, you are apt to get more by investing in government bonds, which is the doctrine of marginal disposability.[218] (2) That, by borrowing the money and paying in the future, you pay by installments, which involves the same kind of productive phenomena {as that much more recently, as I tried to elaborate in my book on the Installment Plan}. People never realized that before. In those two points you have the economic justification for public credit. Of course, you can abuse public credit just as well as private credit or anything else. The most interesting problem is the question of how, through public credit, you are postponing anything into the future.

One reason why nations borrow money, one reason why public credit is legitimate, is just because it is through this process of postponement, which is economically sound, that they are able to meet present obligations. We heard a great deal about it during the war. The question would seem to be, when are they going to pay? The same way as railroads are paying. A railroad would never be built had it not been for this process.

## 113. **Influence of Public Indebtedness**

There are three influences: (1) Political influence. (2) Social influence of public debts. (3) Economic influence. Again, the question whether this influence is going to be salutary or the opposite, depends upon the rate of interest and the condition of the capital market. You may under certain conditions take away capital funds from the capital market which are better employed by private enterprise. Under certain conditions that is best, under other conditions it is not.

## 114. **Debts versus Taxes. Peace and War Finance**

That was largely an academic question. It is now a very great question in the states and cities. New York City, for instance, has gotten along but has heaped up debts of over a billion dollars; Kansas City, also. That problem is now becoming the important one, but the burning question arose during the war. There were two extreme parties at the outbreak of the war: (1) Certain businessmen who wanted us to do what Germany and France had been doing, that is, to pay all the expenses of the war through borrowing. Opposite that came (2) a movement of some academic people who, without thinking deeply over the matter, came out in favor of paying all of the war expenses by taxes. And, of course, the truth lies midway. Adoption of either would cause collapse. The first method would have brought us to the same condition as Germany and France; the second method would have brought the complete collapse of all productivity.

Now, the arguments in a nut shell. The advantages of loans are:

(1) They are voluntary, whereas taxes are compulsory.
(2) They avoid disturbance caused by any certain great change in taxation.
(3) They spread over a longer time, a burden which ought not to be borne by the generation in which the war breaks out; of course, this argument applies to war.

(4) It renders possible far more satisfactory economic adjustments of the burden.

On the other hand, the advantages of taxes are:

(1) That it brings home a public sense of real burden and checks foolish and wasteful extravagance.
(2) If there is not a great deal of disposable capital, taxes may be better than loans.
(3) A pay-as-you-go plan is honest because you don't deceive yourself.
(4) Debts are frequently inadequate, especially in the beginning of the war.
(5) Borrowing can go only to a certain point; unless there is a solid basis of taxation, you get into trouble.

The disadvantage of loans is sometimes said to be producing inflation. That argument, however, is not a very good one, because the same amount of money received by taxation will also cause inflation. On the other hand, the disadvantages of taxes are also great: High taxes will dislocate industry and therefore disturb productivity; primarily during peace time, if the levy is very high, taxation checks consumption, which is the basis of economic prosperity.

Summing up, the conclusion will be as follows: For ordinary expenses in time of peace you want to rely primarily upon taxes. That applies to building parks, school houses and everything else. On the other hand, extraordinary expenses ought to be paid for in general out of loans, but when these extraordinary expenses are expected to last for some time, as, for instance, [in the aftermath of] an earthquake in Japan, then your expenses go on for a long period and you have to adjust your system of taxation to provide a solid basis. So that the real solution of the problem lies in a combination of loan policy and tax policy. The last war is a proof that, while you must rely primarily upon loans, yet if you rely exclusively upon loans, you are going to get into trouble. So that there still remains a very great function for public credit during the war and in time of peace. Governor Smith[219] pointed out recently that the strict application of the pay-as-you-go plan is correct. The strict application in the case of an improvement is that you don't have the improvement because capital is not available through taxation. Strict application of the pay-as-you-go plan is that you are not going to have the automobiles that you have in this country today.

## 115. Classification of Public Debts. The Chief Classification

(1) According to the degree of consent of the lenders, either compulsory or voluntary. Compulsory loans tried in earlier times, 75–100 years ago in Australia and New Zealand, but in the main have been abandoned.

(2) Patriotic gifts or loyalty payments, which means that the government tries to persuade its loyal citizens to accept less than what a certain thing is worth. Nowadays all these things are of very little consequence.

We rely upon a plan of business transactions, contractual loans, instead of periodic loans where they give the government something {that they can take back}. The second criterion is according to the object and the period. We divide into funded loans, ordinary bonds and unfunded loans or floating loans or paper.

We deal now with funded loans because in government, as in private corporations, there are strong arguments against large floating debts. You remember the Atchison Railroad had entirely too large a floating debt, therefore most of the present ones are funded debt. The division into bonds and annuities is made: The government promises to pay the principal at a certain date and the interest. This is the case with bonds. With annuities, you don't get the whole thing at once, you get it every year, a little more than the interest, so that at the end of certain period you get both capital and the interest. There are also life annuities, where a person receives a certain sum for ever [sic]. In a great many other countries, they still have these annuities.

We will devote a little more time today to further classification, where the criterion is the time of the payment and where bonds are classified into either temporary bonds, perpetual debts, or intermediate debts. Temporary debts are what are called floating debts; sometimes they are in form of very short bonds. It is very much like floating debt, and the arguments are just the same. The Secretary of War was telling me that up to the date of the Armistice his great point was to issue bonds for as long a period as possible, inasmuch as no one knew how long the war was going to last. The day of the Armistice, he turned right about and made the period much less.

I want now to talk about perpetual bonds. What is a perpetual bond? A perpetual bond is a bond in which no time is mentioned when payment can be demanded by a creditor. When no time is mentioned it is called a perpetual bond. That is absurdly paradoxical. Why call a bond perpetual? Nothing is said about repayment, and the government can redeem the bond at any time, and yet it is called a perpetual bond. Why? Because if the holder or buyer of the bond is not guaranteed a permanence, if therefore the government does not guarantee permanence, the holder will insist upon government making the government issue a discount bond. What is a discount bond? Suppose government wants to borrow $100,000,000 and the rate is 5%. Now, government can issue $100,000,000 at par, but instead of that issues a 3% bond at 60. As a matter of fact, however, this is of some advantage to the government, because instead of selling at 60, those bonds generally sell at 64 or 65; this way, the government gets a little more. Those countries such as France who don't pay their debts, generally put their bonds in this shape; we don't.

Over against these perpetual and temporary bonds there are intermediate bonds. There are all sorts of them: (a) Ordinary, straight bonds, where all the bonds end at a certain time. (b) A redeemable or callable bond, where government has a right to call the bond before maturity. This is also found in private finance, and, of course, they don't sell as good. (c) A method that is becoming very common in this country is limited option bonds, which are redeemable at one period and callable at another. (d) Serial bonds, and bonds which are issued in such a way that they can be retired in regular installment. We have had great discussions over serial bonds over against straight bonds. If you issue ordinary straight bonds, you have to provide payment for that bond by a sinking fund. You have trouble with sinking funds and therefore in order to obviate the necessity for a sinking fund they devised this method of a sinking fund. Serial bonds have developed now, especially in our states. The State of Massachusetts started some ten years ago and the State of New York followed it. In some states you cannot issue anything except serial bonds.

## 120. **Premium and Discount Bonds**

Here the criterion is the relation between capital and interest. Where you have ordinary high interest bonds and over against them you may have premium bonds, or discount bonds. Premium bonds throw in some premium, as for instance for $100, you get $110. Or in England, throw in some annuities; in France, in the form of lotteries. Discount bonds are very common abroad, but are not advisable for us. The advantages might be put as follows: The advantage to the lender is that he gets the premium and enjoys interest. In the meantime the advantage of the government is that instead of getting $100,000,000 it gets $105,000,000, and gets it easily. They sell easily. The disadvantage lies when the government is disinclined to pay the debt. [Some confusion here.]

## 121. **Contraction of Public Debts. Methods of Emission**

I will mention only three big questions: (1) Should a government place the loans through bankers, or sell them itself? We have tried both plans, and when government credit is wobbling of course you have to employ bankers to do it. During the Civil War, J. Cooke[220] [Cook in original] was able to come to the rescue of the government and sell hundreds of millions worth. If the government can do it by itself, it is better, as we did during the last war. You remember during the war everybody had to take a bond, it was tar and feathers for the one who did not buy his share; of course, this was through the influence of war. (2) Should the government demand a total payment of principle at once? You see, if government takes $100,000,000 from the capital market it might cause a disturbance. It is better to take it a little at a time as you need it. Great mistakes had been done during the Civil War. (3) In borrowing money, should the government pledge any special revenue? That used to be the universal plan, but in modern times this is necessary only in countries whose credit is low. In most cases guarantee is illusory. [?]

## 122. **Conversion of public debts. Refunding**

This matter is of great importance, because most debts have been transacted for war purposes, and during the war the rate of interest goes up, capital is destroyed. For instance, England borrowed at 2% or $2\frac{1}{2}$% before the war and during the war it went up to 6% or 7%. After a few years it becomes very desirable to reduce the interest. Long-term bonds after the war are not desirable because you are going to pay a high rate of interest. If the general rate of interest falls you can perhaps persuade the holders to take the bond at a lower rate of interest; that is called refunding, or converting. In 1877 we put out long-term bonds, then it was reduced to $28\frac{1}{2}$ years. We spent tens of millions of dollars owing to this mistake. One great question is, shall you have frequent conversion, or shall you have few conversions with very great saving? In England they have followed the first principle; we have followed the second plan. The choice depends upon the answer to the question. If you believe in perpetual debt, then the first method of frequent conversion is preferable because it is much more economic; but if, as in this country, we believe [in] paying our debts, because we are able to do so, then the second plan is better, although it is much more extravagant; we are, therefore, justified converting some our large war loans 4%, $4\frac{1}{2}$%, reduced to 3%.

## 123. **Redemption and Payment of Public Debts**

Ought to pay it? The arguments in favor of not paying the debt (I don't mean repudiating the debt. [The issue is paying off the debt]): (1) That the burden diminishes, as the value of money changes. That is true, in the 19th century. Now we know it is not an argument, while it goes down sometimes it goes up some other time. (2) Burden diminishes with the growth of population and wealth. (3) Posterity gets the assets of the present and why should it not get the debts of the present? (4) Want to keep debt for investment for our national banks. Those arguments are really no good. The money argument is better, second argument is not very strong because, although the burden diminishes, yet you must remember that new debts are coming along all the time, and, furthermore, if you are very prudent in your getting rich it is possible to accomplish both, reduce debt and yet be prosperous. The argument of posterity goes too far, because each generation will have its difficulties. The last argument is fallacious because we have better banks without that system now.

So that the general argument would be, while a tax is a burden, yet debt is also a burden, because you must tax yourself to raise the interest on the debt. But how fast you can pay it, it depends. If you can it is always wise to pay off your debt. As a matter of fact France and England had not been able to pay their debts partly because they are too heavy, partly because democratic countries always get enthusiastic. If you are going to pay the debt, what is the use of a sinking fund? A sinking fund really involves a fallacy. It assumes that compound interest would amount to a very great extent. When Pitt adopted that idea, and Hamilton in this country, they assumed that there would be a surplus. But the danger and difficulty is that when emergencies arise the surplus may not exist. Then you have the absurd situation which happened in this country, that in order to go along we had to borrow at 8% and save at 6%. Our present-day sinking fund is different. The law requires that a property tax also be imposed to pay the interest. In the federal government the real thing is to have a surplus.

The problem in the states differs from the problem of the federal government. All state and local debts have been issued for productive purposes; federal debts have been issued almost entirely for war purposes.

Finally, if we had time I would have gone on to show the most disadvantageous way of borrowing money, which is by issue of paper money. If your issue is fiat money, this is nothing but a compulsory tax. In Germany, they virtually took the entire amount of property.

# PART V

# THE BUDGET

## 128. **History of the Budget**

The budget may be defined as being the periodical financial statement containing a report of estimates of revenue and expenditure. This periodical statement is something of a very recent growth. You find something like it in Antiquity. In the Roman Empire you find a distinction between agrarian, general treasury and fiscus, which gradually came to be known as the property of the Emperor himself. And the word fiscus is still employed in a number of European countries, such as in Germany, to represent this idea. In the Middle Ages, potentates had affairs to look after, then small potentates became national monarches. In England you have the early stages, reports were made to the king himself as to his own property. That began in the 15th century, but it was not until after the revolution, until the 17th century, that the recognition was made that the public money belongs to the community rather than the king. There was a great fight with the Stuarts, and the Bill of Rights finally settled that once and for all. Kings now were put on a salary. Gradually in the course of the 18th century a system arose which gave more precise estimates and indications of both revenues and expenditures. It was not until the end of the 18th century that the system was pretty well worked out. Most of the revenues were then turned into the so-called consolidated fund, consolidated of a large number of different revenues. In England today there is still a distinction between supply, surplus and consolidated fund. But, you may say that by the end of the 18th century the Parliamentary control over finance was definitely achieved, and reports regularly made to the government.

In France, potentates at one time promised to have just as much importance as in England. There came a movement that, as it was, generally really did not amount to very much. After the 17th century, the so-called Parlements simply followed the decision of the monarch but had no independent power, they had semblance of power – *lit justice* [?]. It took the revolution to bring about Parliamentary control. In 1789, the National Assembly passed a law that people had the right to vote, but even then it took some time. In fact, it was only after the Prussian war that it was definitely decided that the real power resided in the Lower House, which [who in original] could put its hand on the supply and compel change in Ministry.

In Germany, of course, we all know what took place at the time of Bismark when he made war on Denmark and Austria with the consent of Parliament. It was not really until much later, and not finally until the revolution, that democratic control of the Budget was generally achieved.

In this country, our Constitution says that all bills raising revenue must originate in the House and then, another clause, Section 9, that no money shall be drawn from the Treasury unless in connection with an appropriation made by law. The Constitution also proclaims that a regular statement of account of receipts and expenses shall be published from time to time. Alexander Hamilton followed this and reported from time to time, reporting sometimes once a week, sometimes in two weeks, sometimes in two years. Finally, there is another clause in our Constitution that no appropriation of money for the support of the Army shall be longer than two years.

Perhaps we can best identify our system by comparing it with the two leading extreme opposites. Compare, first, our methods with budgetary affairs in England. The first item is to prepare an estimate. The fiscal year in England starts in April, the 1st of April. Therefore, six months before that date the Department of Treasury sends a circular to every administrative government [agency] inviting their recommendations for the next year. They are all sent to the Chancellor of the Exchequer. He does with them as he likes; he might strike all items. The Chancellor of Exchequer presents it to the Cabinet. The Army and Navy estimates are presented separately in different volumes. They are next considered by the Cabinet. Then, the Cabinet, through the head of the government, presents this to Parliament. Parliament now appoints its committee of supply and these estimates are presented in three volumes: Army, Navy, and Civil Service. The heads of every Department are present, so that when each item is presented they may be questioned, or they may not be. There are from 150 to 200 separate votes. If there is a single vote in opposition to the program [idea in the original] of the head of government, the cabinet at once resigns. That is the common way. In other words, it is vital. The Parliament reserves to itself the appointment of the Committee of Ways and Means, and it is before this committee that the Chancellor of the Exchequer makes his annual budget speech. In this budget speech, he estimates the receipts and figures out a surplus of three to four hundred thousand pounds, never more, never less. He next {introduces reservation toward making a grant}. Then discussion is made. This may be disapproved either in whole or in part, and whether or not the Cabinet resigns depends upon the importance of the question. After it goes through the House [Parliament in the original], it goes to the Lords, who now have very little power over the matter. The Lords cannot amend the bill. The last grant of Ways and Means is one to the appropriation committee. The head of the government has his fingers in from the beginning to the end. That is one extreme.

In our system, the fiscal year begins July 1st, and nine months before that each Department sends the Secretary of the Treasury (in the system which existed until 1921) estimates of what it needs. The Constitution says that the Secretary should report. Hamilton asked, "To whom shall I report?" The custom arose that the Secretary sends his report to the Speaker of the House, who turns it over to different committees. Originally there was only one appropriation committee. We now have thirteen or fourteen, each of which is supposed to present appropriation bills. These appropriations do not include permanent appropriations; such are not included in the annual discussions. The bill is next considered in the House as a whole. Department heads may be called to testify, and any member of the House may propose an amendment. After the committee as a whole reports, the matter is brought before the House and is no longer debated, the House finally accepts it. This then goes to the Senate, which makes changes if it so wishes. Then the question arises, what if one branch believes one way and the other believes the other way? What are you going to do about it? A conference committee was originally designed only to iron out the differences between House and Senate, but the result has been entirely different. The situation is that all sorts of questions are presented to the conference committee. The conference committee finally sends its report to each House. In the meantime we might have all sorts of absurd results. In his annual report the Secretary of the Treasury gives his views on the revenue side of this, and nine times out of ten no one knows the intention of his views. He decides one way, and just because the Secretary of the Treasury decides one way, Congress may decide the other way. After many months it goes to the Senate, to what is called the Finance Committee. It is utterly impossible to make even a gross estimate of budgetary results. The guess is made within a few hundred millions of dollars, so we live in successive surpluses or deficits of a few hundred millions of dollars. That has been our system.

There have been two great changes introduced into this system: one in the beginning of the appropriation process, the other at the end, after the budget has gone through. In the matter of preparation, a director of the budget is appointed who is personally responsible to the President and decides what each department ought to get. That is about as far as he goes. Change has also been introduced in the matter of spending money: We have a general official controller. In between the preparation of the budget and final control over the money expended exists the whole range of matters that I spoke about earlier; no changes have been made.

The difference between the English system and the system that we have is that whereas in England you have centralization and responsibility in the hands of [a few] men, in this country you have neither centralization nor responsibility in the hands of any man, least of all the Secretary of the Treasury. When the executive belongs to one party, the other branch [members in the original] belongs to the other party.

France is interesting because it represents the third kind of budgetary system. In France, the fiscal year end is the fiscal year, January the first to January the first. The trouble there is that in order to get your estimate in time for a discussion you have to start about fifteen months ahead; the English six months, we nine months, and the French fifteen months. They divide the budget into two parts: the budget of revenue and the budget of expenses. The estimates are prepared by the Minister of Finance. They are introduced as a bill prefixed by what they call *expose des motiffe*. This is sometimes a huge volume of about 1500 pages, discussing every problem under the sun. For instance, 50 to 100 pages discuss Allied debt. At all events, the *rapporteur* is generally a distinguished statesman, but also a distinguished economist, and that is the one great thing they have over England and the United States. Here we have nothing, whereas in England you have the speech of the Chancellor of Exchequer. The result of this is a bill that must be introduced in the lower House. In France, the Senate can introduce no finance bill at all; with us, revenue bills are limited to the House. After the budget is printed, it is referred to the budget commission of thirty-three members divided into eleven subcommittees, each of which examines the details. They report to the main committee and the whole matter is discussed in the main committee. The committee has very great powers. They can call for written explanations and then will pass the law. After the bill has passed the House, it goes to the Senate. The question as to whether the Senate has the right to amend was long undecided, and was decided only a few years ago. Once it is passed, the President may not like the budget, and if he does not like it, he may re-submit it to the Parliament. They have to debate it over again, but if they come to the same conclusion, it goes through; anyway the President has to sign it. So you see that in France they have a sort of medium between the British system of complete centralization and responsibility and our American system, so-called committee system and irresponsibility.

## 130. **Preparation, Form and Composition of the Budget**

In England, the executive prepares the entire budget. In this country, now, the Director of the Budget prepares the budget. The budget law of June 10, 1921, requires the President to submit to the Senate estimates which are in his judgment necessary. The President appoints the Director of the Budget to do this task for him, The estimates must all be done by the Director. He confers with the members of the Cabinet. Although he does not belong to any political party, the estimates will be voted by the dominant party. On the whole we have been very fortunate with our Directors of the Budget. When the estimates have been prepared, the budget goes to the different committees {a year or two}. However, after the budget has passed each house, first the House and then the Senate, they form one committee on appropriations, so that we have now very much the same kind of system that France had for a long time. Committees, of course, don't need to and generally don't follow the recommendations of the Secretary, and the only important power that the executive has is the threat of veto. In our states, under the so-called executive budget that has been developed in Europe and other countries, the preparation of the budget is really taken out of hands of various committees and is put in the hands of somebody appointed by the governor, so that we might consider this a progress in the way of preparing better and defensible budgets.

One other question that is involved with the budget is the dates of the fiscal year. Different countries have different dates. We used to have different dates. Up to 1844, we had the calendar year. You find this period in almost every country.

The English system seems to be on the whole about the best; it is always just enough. In France it is most unfortunate.

The next question is the question of unity and completeness of the budget, the question of centralization. We have in our federal and state governments a large number of funds. Instead of having one fund, some have 50, 60, 70, 80 different funds; hospital funds, university funds, so that it is impossible to find out what government is spending; this, of course, is an immense abuse.

The important problem in the matter of budget is whether you ought to have a gross or a net budget. A net budget is the amount of expenditure less the cost of administration. The gross budget comprises all expenditures. In former times, all budgets were net budgets, because money was raised by the local authorities. In modern times, we have a developed state, and central administration; it is different. The advantage of a net budget is that you can know exactly the revenue of the branch of administration. The disadvantage of a net budget is that you never know what you are spending. The advantages of a gross budget are, therefore, economy and clarity. All of our budgets nowadays are gross budgets with one peculiar exception, our post office. That is because the ordinary revenues and expenditures balance and are entirely in the hands of the post office authorities. In other countries, they exclude post offices.

The other side is whether general state budgets should include local budgets, that is to say, should include the money that is paid over to the local units {revenue derived from other kinds}. For instance, take the income tax figures. In our state budgets you will find that the income tax revenue includes 50%, which comes to the state. You would not think that revenue is quite as great as it seems. There are a good many arguments on both sides. For the purpose of clarity here again the gross budget is preferable.

Another question is the method of calculation. How should your estimates be put in preparing your budget? In England they allow a slight surplus, they would do away with certain things. That system is far better than our system where we vary with periodical deficits.

Still another question is, what should be the general arrangement of your budget? It ought to be simple and at the same time not arbitrary. It should cover everything that is important but it should not require every particular detail. {Now budget is prepared and it is prepared by record.} All these have never been discussed by the Director. This is quite different from the French *expose des motiffe*. What is true of the Director of the Budget is still more true of the budgets of different states.

## 131. Presentation, Discussion and Vote on the Budget

Here again we have more important problems: (1) How shall the legislature be organized? [How shall be the legislator? in the original] In England, they have a committee, we have the Committee of Ways and Means. We now have for the last few years a single committee of appropriations. Mr. Madden,[221] who did very excellent work, died a few weeks ago. In most of our states, we don't even have that, therefore no one is responsible for the budget. (2) This is the power of different Houses in discussing the budget. The Lower House everywhere always has the most power in the matter of revenues and bills. We haven't got the time to go into detail, but here you have two different semi-independent bodies. It is complicated in France and also with us. (3) How long do the budgets last? Here we come to a very important question. In no country is it voted every year, but it seems absurd to put to both legislatures every year the question of the salary of the president, interest on debt, etc. Every country now sets aside certain parts of the revenue and expenses which are not open to annual discussion and remain until changed by the House. In England, it is called the consolidated fund. The consolidated fund includes interest, principal of debt, pensions, diplomatic expenses, etc. In France, Mirabeau had the same system introduced but it never was successful. In this country, we make a distinction between our annual and permanent. Annual appropriations call for no mention. Permanent appropriations are such appropriations as private bills where nothing is left to the executive offices for examinations and inquiry except to identify the party. A permanent specific appropriation is one where the amounts are {subject or principles] are specifically designated and generally for permanent terms because it is not limited in time. You might say, this is definite in amount but indefinite in time, otherwise there will be all sorts of embarrassment and dangers involved.

In our city, we have appropriations for the schools that come from Albany which we cannot touch. As a matter of fact, while the budget is large enough, it does not include all the expenses of the government. They don't figure in the annual budget.

The next question is, where you have a fixed duration, should it be every year, every even year, or sometimes still more? The danger of making it too long is that it will become no longer exact. The general tendency is an annual budget.

The other question is the specialization of the budget. Shall you vote on progression, or shall you vote on chapter? Formerly, budgets always were worked on as a whole, the only exception being Greece. Everywhere else you vote by chapter. In France, the number is bigger. We go very much to the extreme. Our budget often requires several thousand votes. In some countries they can transfer from chapter to chapter, but in our federal government this is not possible because our Constitution does not allow the making of transfers by the executive. Therefore, the centralized system again seems the middle course. The English system in this respect is not sensible.

The next question is, whether, after the budget has been voted, you should have any supplementary budgets. Of course, during the war emergency we had supplementary budgets. The idea is to keep this to as low a figure as possible. With us no executive appropriation is possible. This has led to the most absurd results. When General Walker was head of the Census, the money gave out while carrying out the work, and there was no money left to pay the salaries of the clerks. In order to continue, he had to come to New York and borrow money on his own credit to keep the work going.

## 132. **Execution of the Budget**

Here again we have several important problems:

(1) How should your accounts be kept – according to cash receipts and payments over against the second method, your income and expenditure? The cash receipt payment system is, of course, simple, but it is incomplete, because you never get such statements as profit and loss, you cannot get a balance sheet from mere cash receipt and payment accounts. On the other hand, you have expenditure, you have every necessary thing to cover your accounts, because all items are either payable or receivable. We made a little progress in the last few years, but most of our government bodies still adhere to the very primitive cash payment system.

(2) How long should the budget last? Here is another very important question of public accounting. In this country, we have learned the theory of accrual over against cash payment. The definition of accrual is the size or amount or stage of growth of an ordinary income. [Awkward] What you want to do is to balance your assets against your liabilities, you don't want to balance your payments in with the payments out. We have had all sorts of difficulty, because of our inability to separate our total account from current expenses.

(3) What is your machinery for spending and receiving your money? That goes into the whole administrative system. It would be interesting to sketch this if we had time. We have had great improvement in the federal government in the last few years. The Director of the Budget has developed chief coordinators. One department one was buying bedding and another department was selling bedding. We have a much better method now, much better balancing of the budget. We also haven't the time to go into the details of the administrative system. The federal government has made great improvements in this respect in the last few years.

(4) The last question under this heading is, how should the funds be kept? In this country we have had three systems. We started out with government banking, and when state banking was abolished, we tried private banks. When that system broke down, we tried the system of an independent treasury, keeping our money ourselves. Now we have the system of Federal Reserve Banks, whereby the Federal Reserve Banks act as agents of the government. In that, of course, we have introduced a great improvement into the whole matter of fiscal administration.

In most of our states we still have home rule, and all the abuses which have been so great for the past generation or two. Perhaps no field in American budgets requires greater reform than this manner of financial machinery.

## 133. **Control of the Budget**

Here we have three different phases. (1) Administrative control was first worked out in England by the auditor general, not a political man. In this country, we have a very multifarious system. We now have a Comptroller General, so we have, in that respect, made considerable improvement. [(2)] The legislative control is always added to the executive administrative control. We now have a standing committee on public expenditure and also one in the Senate which examines the matter very carefully and makes a report which is widely read. [(3)] In addition to these we have judicial controls, but judicial control has reached its chief development primarily in France and in the

Romanic countries. This goes far back to the Middle Ages, when they were called *Cour des Comptes* [*Corps de Compt* in original].[222] This was composed mostly of eminent judges.

Summing up, I would wish to bring home to you the difference between this country and other countries. The only improvement we have made has been in the preparing of the budget and a slight change in its control. The reason why we are so backward is the deep-seated reason of our political constitution. European governments and other governments have the cabinet system. Our system is based upon the theory of separation of powers, the theory in which we now know an error was made by Montesquieu. That the theory of separation does not exist [in practice], because the executive are state officials as are the legislators, and therefore the executive, being in harmony with the legislature, can keep his fingers on the budget from the beginning to the end. Here government belongs to one party and the legislators to a different party. It is too much to hope that we would have an overhauling of our Constitutional theory, but we are now beginning to realize the fallacy of the whole thing. Our real difficulty lies in our Constitutional difficulty. Nowhere is this more felt than in this budget problem. I hate to close with these pessimistic notes, but, as regards this particular matter, we must remedy the situation.

Hoping that you will all pass a brilliant examination.

# NOTES

1. Henry Carter Adams, Possibly Antoine Lavoisier (1743–1794); best known as a scientist, he was also involved in creating the National Income Accounting System in France. His main economic work "De la Richesse Teritoriale de la France" was unfinished at the time of his death – the main part of that work, however, was published as Melang'es *d'Economie Politique* in 1847. 1851–1921; *The Science of Finance*, New York, 1898.

2. Adolph Heinrich Gotthelf Wagner, 1835–1917.

3. Laws of Menu [Manu]: Indian sage and lawgiver; a collection of the earliest laws of ancient India – of very remote antiquity (between 1200 and 1500 B.C.). "Manu's Maxim" is given by Plehn in terms of tax burdens pressing within equal severity on each individual, but not readily in terms of mere numerical proportion. Carl C. Plehn, *Introduction to Public Finance* (4th ed.). New York: Macmillan, 1921, p. 4.

4. Kautilya (Kautalya), *Arthasastra.*

5. Narendranath Nath Law, *Studies in Ancient Hindu Polity, Based on the Arthasastra of Kautilya*, London: Longmans Green, 1914.

6. Chuang Tzu (369–286BC), a follower of Lao Tzu (a contempory of Confucius, but in the Taoist tradition). Chuang Tzu took Lao Tzu's philosophy on laissez faire to the extreme, advocating individual anarchism.

7. Xenophon, 434?–355? B.C.

8. Aristotle, 384–322 B.C.

9. Likely Pliny the Elder, 23–79, who blamed the Latifundia for the downfall of Rome and dealt with taxation issues; possibly his nephew, Pliny the Younger, 62?–113?. Yet neither Pliny wrote the Book of Census – Pliny the Elder's works were in history and natural science and Pliny the Younger published his voluminious correspondance and speeches.

10. Cassius Dio Cocceianus, ca. 150–235. *Dio's Roman History*, Earnest Cary, trans. London: W. Heinemann; New York: Macmillan, 1914–27.

11. Ibn-Khaldun, 1332–1406; *The Muqaaddimah: An Introduction to History.*

12. Started in 1081, completed in 1086, by order of William the Conqueror.

13. *Dialogus de Scaccarrio*, or *Dialogue of Exchequer*, by Richard Fitzneale (also Fitznigel), 1130–1198, treasurer of England under Henry II and Richard I.

14. Diomeda Conte di Maddaloni Caraffa, 1406–1487.

15. *Speculum*: Title of several of the most ancient lawbooks or compilations. An ancient Icelandic book is called *Speculum Regale*.

16. Sir John Fortescue, 1397?–1476?; *The Governance of England: Otherwise Called the Difference Between an Absolute and a Limited Monarchy*, 1471 (Oxford, 1885).

17. Possibly Raffaele Ciasca, *L'arte dei Medici e speziali nella stieria e nel commercio florentino dal secolo XII al SV*, Firenze: L. S. Olschki, 1927; or Graziella

Silli, *Una Corte all fire del 500: artisti, letterati, scienziati nella reggia di Ferdinando I de Medici*, Firenze: Fratelli Alinari, 1927.

18. Niccolò Machiavelli, 1469–1527.

19. Francesco Guicciardini, 1483–1540; *Avvertimenti politici*, 1583.

20. Martin Luther, 1483–1546.

21. Desiderius Erasmus, 1466?–1536.

22. Melchior Von Osse 1506–1557; *Testament gegen Hertzog Augusto, Churfürsten zu Sachsen*, Halle, 1556.

23. Jean Bodin, 1530?–1596; *Les six Livres de la Republique*, Paris, 1577; *Discours sur le rehausssement et la diminution des Monnaies*, 1578.

24. Giovanni Botero, 1540/1544–1617; *Della Ragion di Stato Libri Diece*, 1590; *Della Causes Della Grandezza Delle Città*, 1588/1635.

25. John Hales, ?–1571; *A Discourse of the Common Weal of this Realm of England*, 1581.

26. Thomas Hobbes, 1588–1679; *Leviathan*, 1651.

27. Sir Thomas Mun, 1571–1641; *A Discourse of Trade from England into the East Indies*, 1621; *England's Treasure by Foraign Trade*, 1664.

28. Sir William Petty, 1623–1687; *A Treatise of Taxes and Contributions*, London: 1662/1667.

29. John Locke, 1832–1704.

30. Charles Davenant, 1656–1714; see his *An Essay upon Ways and Means of Supplying the War*, 1695, *Discourses on the Publick Revenues*, 1698, and *An Essay upon the Probable Means*, 1699.

31. Cardinal Jules Mazarin, 1602–1661.

32. Sébastien Le Prestre, Seigneur de Vauban (Marshall Vauban), 1633–1707; *An Essay for a General Tax, or a Project for a Royal Tithe*, London: 1709.

33. Pierre le Pesant, Sieur de Boisguilbert (or Boisguillebert), 1646–1714; see his *Détail de la France*, Paris, 1697.

34. *Le Détail de la France*, 1696.

35. *Projet de dime royale*, 1707.

36. Presumably Christoph Besold, 1577–1683.

37. Unknown. Bruno Mallù's text *Zur Geschichte der Vermogensteuern um Mittelalter* was published in 1911 in Leipzig. This is cited in Seligman's contribution to *Essays in Taxation* and is in the context of local taxation in Germany. There is also a reference to his *Zur Geschichte der englischen und amerikanischen Vermogensteuern*, 1912, in the same essay. But there is also Otto Muller, who wrote *Die Einkommenbesteuerung in den verschiedenen Landern*, published out of Halle in 1902; and Otto Maull who wrote in political geography, but no other reference to him seems to have been made in Seligman's works.

38. Jakob Bornitz, 1570?–1630.

39. Presumably Hermann Latherus, 1583–1640; *De Censu*, Frankfurt, 1618.

40. Unidentified.

41. Kasper Klock, 1584–1655.

42. Velt Ludwig von Seckendorff, 1626–1692.

43. Hermann Conring, 1606–1681; one of first lecturers on public administration in Germany; author of *Staatsbeschreibung* (1660) and *De Vectigalibus et Aerario* (1663).

44. Presumably Johann Joachim Becher, 1635?–1682.

45. Freiher Wilhelm von Schröder, 1640–1688.

46. Presumably Kurt Zielenziger, 1890-, *Die alten deutschen Kameralisten Ein Bietrag zur Geschichte der Nationalökonomie und zum Problem des Merkantilismus*, Jena: G. Fischer, 1914; and *Gerhart Von Schulze Gaevernitz; eine darstellung seines wirkens und seiner werke*, Berlin: R. L. Prager, 1926.

47. The "end of the 17th century" seems out of place here; possibly Arthur Sommer, 1889-, *Friedrich List's Sytem der Politischen konomie*, Jena: G. Fischer, 1927.

48. Karl Heinrich Rau, 1792–1870; *Grundsätze der Finanzwissenschaft* (2 vols., 3rd ed.). Heidelberg: C. F. Winter, 1850–1851.

49. Julius Bernhard von Rohr, (1683–1742), a protestant cannon in Mecklenberg interested in economic questions. In the thesis for his doctor's degree he took up the subject of the advancement of economic study and argued for the establishment of professorships in economics. Later he devoted efforts to the establishment of subsidies for economic societies. His dissertation was "De Excolendo Studio Oecono tam Principum Quam Privatorum", published in Leipzig, 1712.

50. Johann Heinrich Gottlob von Justi, 1717–1771.

51. Joseph von Sonnenfels, 1733–1817; *Einleitungsrede in Seine Akademische Vorlesungen*, 1763; *Grundsätze der Polizei*, Vienna and Munich, 1781.

52. Jacob Vanderlint, ?–1740; *Money answers all Things; or, an Essay to make Money sufficiently Plentiful, amongst all Ranks of People, and Increase our Foreign Trade and Domestick Trade, Fill the Empty Houses with Inhabitants, Encourage the Marriage State, Lessen the Number of Hawkers and Pedlars, and in a great measure, prevent giving land Credit, and making bad Debts in Trade*, London, 1734.

53. Comte Henri de Boulainvilliers, 1658–1722; *Etat de las France*, 1766; *Des états Généraux, Et Autres Assemblées Nationales*, 1788–1789

54. Abbé Charles-Irénée Caste; de Saint-Pierre, 1658–1743.

55. Marquis d'Argenson, 1694–1757.

56. Jean François Melon, 1675–1738; *Delle Monete Controversia Agitara Tra Due Celebri Scittori Oltramontani*, 1754; *Essay Politique* Possibly Antoine Lavoisier (1743–1794); best known as a scientist, he was also involved in creating the National Income Accounting System in France. His main economic work "De la Richesse Teritoriale de la France" was unfinished at the time of his death – the main part of that work, however, was published as Melang'es *d'Economie Politique* in 1847.

57. Charles de Secondat, Baron de la Brède et de Montesquieu, 1689–1755.

58. Claude Dupin, 1684–1769.

59. Richard des Glanieres (Glannieres); *Plan d'imposition économique et d'adminis-traion des finances*, Paris, 1774.

60. François Quesnay, 1694–1774.

61. Victor Riquetti, Marquis de Mirabeau, 1715–1789.

62. Pierre François Mercier de la Riviére, 1720–1793.

63. Pierre Samuel Dupont de Nemours, 1739–1817.

64. Abbé Nicolas Baudeau, 1730–1792; see his *Lettres d'un Citoyen à un Magistrat sur les Vingtièmes et les autres Impots*, Paris, 1768.

65. Guillaume François (Francis) Le Trosne, 1728–1780.

66. Anne Robert Jacques Turgot, 1727–1783.

67. Jean Joseph Louis Graslin, 1727–1790; see his *Essai Analytique sur la Richesse et sur l'Impt*, London, 1767.

68. Moreau de Beaumont, 1715–1785. Smith is referring to *Memoires concernant les Impositions et Droits en Europe*, 4 volumes, Paris: De l'imprimerie royale,

1768–1769 [What is listed is the title of the first volume. The remaining volumes were entitled *Memoires concernant les Impositions et Droits, 2de. Ptie., Impositions et Droits en France*.]; 5 volumes, Paris: Chez J. Ch. Desaint, 1787. Apparently, Smith acquired his copy through Turgot.

69. Pierre-François Boncerf, 1745–1794; *La plus importante et la plus pressante affaire, ou, La nécessité et les moyens de restaurer l'agriculture et le commerce*, 1790; *De la nécessité et des moyens d'occuper avantageusement tout les gros ouvriers*, Paris, 1791.

70. François Marie Arouët de Voltaire, 1694–1778; L'Homme aux quarante écus, 1768.

71. Jean-Jacques Rousseau, 1712–1778.

72. Jacques Necker, 1732–1804.

73. Jacques Pierre Brissot de Warville, 1754–1793. See note 77, *infra*. Possibly John Briscot, 1652–1728.

74. Marie Jean Antoine Nicolas de Caritat, Marquis de Condorcet, 1743–1794.

75. Abbe Sieyès: Emmanuel Joseph Comte Sieyès (1748–1836).

76. Etienne Claviere (1754–1793) was a minister of finance during the same time as Roland [next note] and worked for the Revolutionary Tribunal, but was nonetheless eventually sentenced to death. Co-author with Brissot of *The Commerce of America with Europe*, London, 1794.

77. Presumably Jean-Marie Roland de la Pltière, 1734–1793, minister of the interior during the early 1790s and executed in 1793, rather than Barthélémy-Gabriel Roland de Fleury (1715–1810). (Pltière and Claviere are frequently mentioned together in history and biography books.

78. Antoine Lavoisier (1743–1794); best known as a scientist, he was also involved in creating the National Income Accounting System in France. His main economic work "De la Richesse Teritoriale de la France" was unfinished at the time of his death – the main part of that work, however, was published as Melang'es *d'Economie Politique* in 1847.

79. Honoré Gabriel Victor Riquetti, Marquis de Mirabeau, 1749–1791.

80. Lione Pascoli, 1674–1744; *Testamento politico d'un accademico fiorentino*, 1733.

81. Sallustio Antonio Bandini, 1677–1760; *Discorso Economico* (also published as *Discorso sopra la Maremma di Siena*), Firenze: Cambiagi, 1775.

82. Carlo Antonio Broggia, 1683?–1763; *Trattao de Tributi, Delle Monete, E Del Governo Politico Della Sanità, Opera Di Stato, E Di Commercio*, 1743; *Del Pubblico Interesse Economico, Politico, Morali, di Stato Edi Commercio*, 1750.

83. Francis Hutcheson, 1694–1746.

84. Sir Robert Walpole, 1676–1745.

85. Jonathan Swift, 1667–1745.

86. Adam Smith 1723–1790.

87. David Hume, 1711–1776; see his *Political Discourses*, 1752.

88. Adam Dickson, 1721–1776; *An Essay on the Causes of the Present High Price of Provisions: As Connected with Luxury, Currency, Taxes, and National Debt*, London: 1773.

89. Sir James Denham Steuart, 1712–1780.

90. John Craig, 1780–1850; *Remarks on Some Fundamental Doctrines in Political Economy*, 1821.

91. David Ricardo, 1772–1823.

92. John Ramsey McCulloch, 1789–1864.

93. John Stuart Mill, 1806–1873; *Principles of Political Economy with Some of Their Applications to Social Philosophy*, 1848.

94. William Stanley Jevons, 1835–1882.

95. Charles Francis Bastable, 1855–1945; *Public Finance*, 1892, 3rd. ed., 1903.

Possibly Antoine Lavoisier (1743–1794); best known as a scientist, he was also involved in creating the National Income Accounting System in France. His main economic work "De la Richesse Teritoriale de la France" was unfinished at the time of his death – the main part of that work, however, was published as Melang'es *d'Economie Politique* in 1847.

96. Edwin Cannan, 1861–1935; *History of Local Rates in England*, 1896, )2nd ed.), 1912.

97. George Bernard Shaw, 1856–1950; *The Common Sense of Municipal Trading*, Westminster: A. Constable, 1904 [on municipal ownership]; see also *The Intelligent Woman's Guide to Socialism and Capitalism*, New York: Brentano's, 1928.

98. Nicolas-François Canard, 1750?–1833; *Principes d'Économie Politique*, Paris: F. Buisson, 1801.

99. Louis Adolphe Thiers, 1797–1877; *De la Propriété*, Paris, 1848.

100. Karl Heinrich Rau, 1792–1870.

101. Presumably Vorgetragen Johann Gottfried Hoffman, 1765–1847, *Die lehre von den Steuern als Anleitung su gründlichen Urtheilen über das Steuerwesen*, Berlin, 1840; and *Geschichte der Direkten Steuern in Baiern vom Ende des XIII Jarhunderts* (n.d.). Had long and distinguished career as a teacher, in public administraion, and as professor of practical philosophy and cameral science at Koningsberg; became professor of political economy at Berlin in 1810.

102. Adolf Held, 1844–1880, *Die Einkommensteuren Finanzwissenschaftlichte Studien zur reform der directen steuren in Deutschland*, Bonn: A. Marcus, 1872; *Sozialmus, Sozialdemokratie und Sozialpolitik*, Leipzig: Duncker and Humblot, 1878; *Zwei Bucher Zur Socialen Geschichte Englands*, Leipzig: Duncker and Humblot, 1881. See also *Carey's Socialwissenschaft und das Merkantilsystem, eine Literatergeschichtliche Parallele*, Wurzburg: F. E. Thein, 1866.

103. Adolf Heinrich Gotthilf Wagner, 1835–1917.

104. Wilhelm Georg Friedrich Roscher, 1817–1894; *System der Volkswirtschaft*, five vols., 1854–1894, *System der Finanzwissenschaft*, Stuttgart, 1886.

105. Gustav Cohn, 1840–1919; *The Science of Finance*, 1895.

106. Friedrich Julius von Neumann, 1835–1910, *Grundlagen*, 1889 [?];*Die Steuer Nach der Steuerfähigkeit; ein Beitrag zur Kritik und Geschichte der Lehren von der Besteuerung*, Jena: G. Fischer, 1880 and *Jahrbücher für Nationalokonomie und Staatswissenschaft*, Bd. 2, 1881; *Die Steuer und das Offentliche Interesse. Eine Untersuchung über das Wesen der Steuer und die Gliederung der Staats und Gemeinde-Einnahmen*, Leipzig: Duncker and Humblot, 1887; *Vermogensteuern und Wertzurvachsteuern als Erganzung der Einkommensteuer, Insbesondere in Wurttember*, Tubingen: H. Laupp, 1910.

107. Robert Meyer, 1855–1914; *Principien der gerechten Besteuerung*, 1884; *Das Wesen des Einkommens*, 1887.

108. Gustav Friedrich von Schönberg, 1839–1908, *Handbuch der Politischen Oekonomie*, Tubingen: H. Laupp, 1882 [same title in three-volume series published in

1890–1891 and again in 1896–1898, with volumes 1 and 2 titled *Volkwirtschaftlehre* and volume 3, *Finanzwissenschaft und Verwaltunglehre*].

109. Emil Sax, 1843–1927.

110. Luigi Cossa, 1831–1896.

111. Guiseppe Ricca-Salerno, 1849–1912, *Le Dottrine Finanziarie in Ingliterra tra la fine del secolo XVII e la prima meta del XVIII*, Bologna, 1888; see also his *Del metodo in economia politica*, 1878; *Storia delle doctrine finanziarie in Italia*, 1881; *Manuele di scienza delle finanze*, 1888..

112. Maffeo Pantaleoni, 1857–1924.

113. Augusto Graziani, 1865–1944; *Instituzioni di Scienza delle Finanze*, Turin, 1897 (2nd ed.), Turin, 1911.

114. David Ames Wells, 1828–1898; see his *The Theory and Practice of Taxation*, New York, 1900.

115. Presumably Francis Amasa Walker, 1840–1897.

116. Richard Theodore Ely, 1854–1943; with J. H. Finley, *Taxation in American States and Cities*, 1888.

117. Henry Carter Adams, 1851–1921; *Public Debts*, 1887; *Science of Finance*, 1898.

118. Winthrop More Daniels, 1867–1944, *The Elements of Public Finance*, New York: H. Holt, 1899; republished 1904.

119. Carl C. Plehn, 1867–1945; *Introduction to Public Finance*, New York: Macmillan, 1896 (5th ed.), 1926 (6th ed.), 1929; *The Tariff Relations of the United States and the Philippine Islands*, Philadelphia, 1904; *Digest of State Laws Relating to Taxation and Revenue*, 1922; *Government Finance in the United States*, 1915.

120. Max von Heckel, 1865–1913, *Die Einkommensteuer und die Schuldzinsen*, Leipzig: C. F. Winter, 1890; *Das Budget*, Leipzig: C. L. Hirschfeld, 1898; and *Lehrbuch der Finanzwissenschaft*, 2 volumes, Leipzig: C. L. Hirschfeld, 1907, 1911.

121. Presumably Walther Lotz, 1865–, *Finanzwissenschaft*, Tubingen: Mohr, 1917 (2nd ed.), 1931.

122. Georg Schanz, 1853–1931, editor, *Finanz-Archiv*, Stuttgart and Tubingen, 1884–. Schanz was very prolific. The following references to him are found in Seligman's other works: *Die direkten Steurn Hessens und deren neueste Reform*, *Finanzarchiv*, vol. xiii, 1885; *Zur Frage der Steuerpflicht*, *Finanzarchiv*, vol. ix, 1892; *Der Einkommensbegriff und die Einkommensteuergesetze*, *Finanzarchiv*, vol. 13, 1886; *Die Doppelbesteuerung und der Volkerbund*, *Finanzarchiv*, vol. 40, 1923, pp. 353–370.

123. Gaston Paul Amédée Jèze, 1869–1953; *The War Finance of France*, New Haven: Yale University Press, 1927.

124. Edgard Allix, 1874–1938, presumably *Les Conributions Indirectes; Traité Théorique et Pratique*, 2 vols., Paris, 1929.

125. John Maynard Keynes, 1883–1946.

126. Arthur Cecil Pigou, 1877–1959.

127. William Robert Scott, 1868–1940.

128. John Atkinson Hobson, 1858–1940; *Taxation in the New State*, 1919.

129. Henry Higgs, 1864–1940.

130. Edward Hugh Dalton, 1887–1962; *Principles of Public Finance*, 1923.

131. Josiah Charles Stamp, 1880–1941; *The Fundamental Principles of Taxation*, 1921; *Wealth and Taxable Capacity*, 1922.

132. G[eorge]. Findlay Shirras, 1885–1955, *Indian Public Finance and Banking*, London: Macmillan, 1919, republished 1920; *The Science of Public Finance*, London: Macmillan, 1924, republished 1925 and, completely revised and rewritten, 1936.

133. Francesco Saverio Nitti, 1868–1953; *Principes de Science des Finances*, Paris, 1904.

134. Luigi Einaudi, 1874–1961.

135. John Gustaf Knut Wicksell, 1851–1926.

136. Nicholas Gerard Pierson, 1839–1909.

137. Charles Jesse Bullock, 1869–1941.

138. Thomas Sewall Adams, 1873–1933.

139. Jacob Henry Hollander, 1871–1940.

140. Harley Leist Lutz, 1882–1975; *Public Finance*, New York: D. Appleton, 1924.

141. Presumably Robert Murray Haig, 1887–1953.

142. Fred Rogers Fairchild, 1877–1966; *The Economic Problem of Forest Taxation*, New Haven, 1909.

143. John Edward Brindley, 1878-, *History of Taxation in Iowa, 1910–1920*, Iowa City: State Historical Society of Iowa, 1921; *Tax Administration in Iowa*, Iowa City: State Historical Society of Iowa, 1912; with Thomas Radford Agg, *Highway Administration and Finance*, New York, 1927; with Grace Stone McClure Zorbaugh, *The Tax System of Iowa*, Ames, 1929.

144. Roy Gillispie Blakey, 1880-; *Effects of Bonds and Taxation in War Finance*, 1917.

145. Harvey Whitefield Peck, 1879-; *Taxation and Welfare*, New York: Macmillan, 1925.

146. A paraphrase of first three paragraphs of the main text of Sir James Steuart, *An Inquiry into the Principles of Political Oeconomy*, Edinburgh and London: Oliver & Boyd, 1966 (1767; 1805), vol. I, pp. 15–16.

147. Nassau William Senior, 1790–1864.

148. Frederich List, 1789–1846.

149. Carl August Dietzel, 1829–1884; *Das System der Staatsanleihen im Zusammenhang der Volkswirtschaft betrachtet*, Heidelberg, 1855.

150. Emile Sax, 1845–1927.

151. Tela: cloth in process of being woven on the loom; secondary meaning: plan or design.

152. In English law, a customary tribute of goods and chattels payable to the lord of the fee on the decease of the owner of the land. This feudal right grew out of the custom under which the lord lent horses and armour to those who served him in battle; when the tenant died, these were to be returned to the lord.

153. *Seisin* was, in English law, the completion of feudal investiture in the tenant. *Primer Seisin*: the right of the king, when any of his tenants died seised of a knight's fee, to receive of the heir, provided he were of full age, one whole year's profits of the lands, if they were in immediate possession; and half a year's profits, if the lands were in reversion, expectant of an estate for life.

154. In old English law, a fine paid to the lords of some manors, on the marriage of tenants, originally given in consideration of the lord's relinquishing his customary right of lying the first night with the bride of a tenant.

155. George Black, *The History of Municipal Ownership of Land on Manhattan Island, to the Beginning of the Land Sales by the Commissioners of the Sinking Fund*

*in 1844*, New York: Columbia University, 1897. *Columbia University Studies in History and Economics*, vol. 1, no. 3; 83pp.

156. The Timber Culture Act of 1873 allowed individuals to claim an additional 160 acres, over the original 160 of the Homestead Act, if they agreed to plant one-quarter of it with trees. The Timber and Stone Act of 1878 allowed people or companies to claim quarter sections of forest land for $2.50 an acre if the land was unfit for cultivation.

157. Gifford Pinchot, 1863–1946.

158. Presumably Sir Horace Plunkett, 1854–1932, Irish reformer of farm life, including that of women; founder and controlling force of the Agricultural Organisation Society of Ireland; with James R. Garfield (1865–1950, son of President James A. Garfield and Secretary of the interior under Theodore Roosevelt) and Pinchot (advisors to Roosevelt) formed the Commission on Country Life in the U.S. to promote farmer welfare.

159. Unable to identify; not a member of U.S. Congress.

160. Andrew Joseph Volstead, 1860–1946.

161. Thomas Jefferson, 1743–1826.

162. Alexander Hamilton, 1755–1804.

163. *Tontine*: In French law, a form of partnership among persons who are in receipt of perpetual or life annuities, having agreed that the shares or annuities of those who die shall accrue to the survivors. Named after Tonti, an Italian who invented the plan in the seventeenth century. The principle is used in some forms of life insurance.

164. "A Lottery therefore is properly a Tax upon unfortunate conceited fools; . . ." William Petty, *A Treatise of Taxes & Contributions*, 1662; in Charles Henry Hull, ed., *The Economic Writings of Sir William Petty*, 2 vols., Cambridge: Cambridge University Press, 1899, vol. 1, p. 64.

165. "No two characters seem more inconsistent than those of trader and sovereign". Adam Smith, *The Wealth of Nations*, Book V, Ch. II, Part I, Para. 7; New York: Modern Library, 1937, p. 771. Smith is comparing the performance of the English East India company as trader and sovereign, saying that if its trading spirit "renders them very bad sovereigns[,] the spirit of sovereignty seems to have rendered them equally bad traders".

166. René Gonnard, 1874–1966, *Histoire des Doctrines Economiques*, Paris: Nouvelle Librarie Nationale, 1921–1922, 1924–1927, 1930.

167. Owen D. Young, 1874–1962.

168. Reed Smoot, 1862–1941.

169. Smith, *Wealth of Nations*, *op. cit.*, pp. 777–778. The reader is invited to compare Seligman's treatments of public expenditures and taxation with those of Smith, as to compass, structure and content.

170. Francis Ysidro Edgeworth, 1845–1926.

171. Thomas Nixon Carver, 1865–1961.

172. Francis Amasa Walker, President of the American Economic Association, 1886–1892.

173. Irving Fisher, 1867–1947; *The Nature of Capital and Income*, 1906.

174. Possibly Gyan Chand, 1893-; *The Financial System of India*, London: Kegan Paul, Trench, Trübner, 1926.

175. Presumably Robert Murray Haig, 1887–1953; "The Concept of Income", in Haig, ed., *The Federal Income Tax*, New York, 1921, pp. 1–28.

176. Likely Frank W. Taussig, *The Tariff History of the United States*, 7th edition revised, New York: G. P. Putnam's Sons, 1923; possibly, Charles F. Bastable, *The Commerce of Nations*, London: Methuen, 1912.

177. Presumably Charles Henry Hull, 1864–1936.

178. Thomas Sewall Adams, "The Effect of Income and Inheritance Taxes on the Distribution of Wealth", *American Economic Review*, vol 5, March 1915, pp. 234–244; "Fundamental Problems of Federal Income Taxation", *Quarterly Journal of Economics*, vol. 35, August 1921, pp. 527–556; "Ideals and idealism in Taxation", *American Economic Review*, vol. 18, March 1928, pp. 1–8.

179. Gabelle: tax on salt. If Seligman's intent here was to indicate something of the circumstances of Turgot's termination as Minister of Finance in 1776, it failed to register. The tax on salt existed in France from the early 1600's until the French Revolution. The Gabelle was an indirect tax which obliged all individuals to purchase a certain quantity of salt for household use regardless of whether they actually needed it. One-fifth of the price was put aside for the use of the state. Furthermore, the salt was purchased at a monopoly price, salt being stored in government warehouses. The tax/price varied greatly from region to region, there was massive fraud by state and tax collecting officials, and it was despised by the general population. The Gabelle provided one of the largest elements of royal income but the state never drew any great direct profit because of the number of officials necessary to administer and enforce the tax. One of Turgot's reforms was a repeal/reduction of the Gabelle as one way to attempt to quiet peasant rumblings. Louis XVI dismissed Turgot because of his economic reform policies; the direct order, however, came from the former minister, Bertin.

180. Chief Justice John Marshall, 1755–1835.

181. Presumably *Essays in Taxation*, 1928.

182. Probably John Jacob Astor (1886–1971), 1st Baron Astor of Hever, second son of William Waldorf Astor (1848–1919), grandson of the fur trader and financier, John Jacob Astor (1763–1848).

183. Voltaire, *The Man with Forty Crowns*, 1768. [See, for example, Constance Rowe, *Voltaire and the State*, New York: Columbia University Press, 1955, p. 158.] An essay on the single tax.

184. According to *Black's Law Dictionary*, Henry Campell Black, ed., Revised Fourth Edition, St. Paul, MN: West Publishing, p. 861: In old English law, a measure of the land that could be worked with one plow; alternately, as much land as would support one family or the dwellers in a mansion house. A unit of taxation, the tax called *hidegild*. [Many of the legal definitions given herein are based on *Black's*.]

185. Predial Servitude: A charge laid on an estate for the use of another estate belonging to another owner.

186. *Vectigalia*: In Roman law, customs duties, or taxes paid upon import or export of certain kinds of merchandise. Differed from tribute, which was a tax paid by each individual. Also, rent from state lands. *Vectigal Judiciarium*: Fines paid to the crown to defray the expenses of maintaining courts of justice. *Black's*, *op. cit.*, p. 1724.

187. Unable to identify. No records of him in New York Times Personal Name Index and Obituary Files, Encyclopedia of New Yorkers, New York Public Library, and New York City Historical Society. (Also Kirby Lowson.)

188. John A. Zangerle, 1866–1956, Cuyohoga County Auditor from 1913–1951, first joined the Quadrennial Board of Assessors in 1913; also had a private law practice.

Author of *Principles of Real Estate Appraising*, Cleveland: McMichael Publishing Organization, 1924.

189. Kai Chau (more often Kiaochow or Kiaochau). The city was occupied by Germany in 1898; the Germans compelled China to sign a 99 year lease in exchange for building three railroads and a port. There were several attempted rebellions before Japan seized the city in 1914. The city was eventually returned to China in 1922. The city was a fairly standard example in the literature of land taxation. Henry George, Henry Farnam, and Richard T. Ely all make mention of it, the reason being that the Germans imposed a particular sort of land tax. Forseeing a dramatic increase in the value of land after the infrastructure improvements, the Admiral of the colony decreed that any transfer of land must be authorized by the government. (The German government originally wanted to buy up the land on speculation and resell it after the improvements but could not afford to do so). In addition, whenever a plot of land was sold, the government was entitled to one-third of the increase in the value of the land, after deductions for improvement. If the plot was not sold, the land was to be revalued every 25 years and the one-third paid to the government. The tax never raised much money, but it did keep speculation to a minimum, especially since anyone authorized to buy land had to do so with the requirement that they build immediately on it. Two articles that address this tax are: W. Schrammler, "Die Landpolitik im Kiauchougebeit" (1911) and "Die Steuer-politik im Kiauchougebeit" (1912) in *Jahrbuch der Bodenreform*, vol. xvii and xviii. The modern city is Tsingtao (also Quingdao) – where Chinese beer is made).

190. Unable to identify. Robert J. Walker (1801–1869) was Secretary of the Treasury, 1845–1849.

191. *Capitatia humana*, likely imposed by Caracalla (211–217) who extended citizenship to many living in Roman lands so that he could tax them more. The *Capitatia humana* (or *capitatia plebia*) was a head tax on serfs (*coloni*), but was really a property tax since it was paid by the landowner.

192. *Nota captivitatis* was the Roman disparaging term for a head tax in Roman times that was applied only to conquered enemies.

193. William Pitt, 1759–1806, British Prime Minister, 1783–1801, 1804–1806.

194. Robert Peel, 1788–1850, British Prime Minister, 1834–1835, 1841–1846.

195. William Ewart Gladstone, 1809–1898, British Prime Minister, four times Prime Minister.

196. A variant of the taille (head tax) that was levied on land rather than on the person; this made it less abusive since there was at least some known basis for the collection. However, it was still only non-noble land that was taxed. It was imposed on some regions during the reign of Henry II. The tax replaced the system of personal taxes that existed before the French Revolution. This was actually a system of semi-personal taxes on land, business, doors, windows, and house rentals.

197. *Impot global*, an "entire tax", essentialy an income tax that is subject to the usual conditions of allowance, exemption, and graduation found in England and the U.S. In this case, the source of income, however, is immaterial.

198. Withholding commenced during World War 2.

199. Possibly of the many death taxes, such as the mortarius.

200. Andrew Carnegie, 1835–1919.

201. Andrew William Mellon, 1855–1937.

202. Seligman should have known better than to make such a prediction. But superior expertise can readily engender such confidence.

203. Sydney Smith, 1771–1845, English clergyman, essayist and wit.

204. David Lloyd George, 1863–1945, British Prime Minister, 1916–1922.

205. This likely explains the absence of sections explicitly numbered 91–97.

206. Seligman, *The Economics of Installment Selling*, New York: Harper & Bros., 1927. Seligman was an important figure in the formation/adoption of consumer installment credit, which he considered vital to enabling mass consumption through payment for items purchased while using them. See Lendol Calder, *Financing the American Dream: A Cultural History of Consumer Credit* (Princeton, NJ: Princeton University Press, 1999.

207. Likely, a general or universal draft, not completely voluntary.

208. Robert Peel, 1788–1850, British statesman, Prime Minister 1834–1835, 1841–1846.

209. The reference here is to perpetual, at least continuing, refinancing of debt.

210. Originally in the 13th century in Genoa, when investors were able to purchase loca, or shares, in the construction of a ship; later the term came to refer to buying shares in joint-stock companies. These loca could be bought short or long term.

211. Presumably Adolph Heinrich Gotthelf Wagner, 1835–1917, *Finanzwissenschaft*, 4 vols., 1871–1872.

212. The figures for the end-of-year interest-bearing national debt given in Paul Studenski and Herman E. Krooss, *Financial History of the United States* (2nd ed.). New York: McGraw-Hill, 1963, (pp. 54, 77, 100, 125, 152, 215, 236, 297, 313) vary somewhat from those recorded in the notes (millions of dollars):

1789–91: 77.2
1812: 56.0
1815: 127.3
1835: 0.3
1846: 16.8
1857: 28.7
1860: 64.8
1865: 2,217.7
1893: 585.0
1899: 1,046.0
1916: 971.6
1919: 25,234.5
1928: 17,318.0

213. John Law, 1671–1729, Scottish financier.

214. Isaac de Pinto, 1715–1787, *Traité de la Circulation et du Crédit*, Amsterdam: M. M. Rey, 1771.

215. Sir John Sinclair, 1754–1835, *Observations on the Report of the Bullion Committee*, London, 1810; see his *History of the Public Revenue of the British Empire* (3rd ed.). London, 1804.

216. Richard Price, 1723–1791, *An Appeal to the Public, on the Subject of the National Debt*, London, 1774, and *The State of the Public Debts and Finances at Signing the Preliminary Articles of Peace in January 1783 with a Plan for Raising Money by Public Loans*, London, 1783.

217. Presumably Carl Friedrich Nebenius, 1784 or 1875–1857, *Der deutsche Zollverein, sein System und seine Zukunft*, Carlsruhe, 1835; *Der öffentliche Credit*, Carlsruhe and Baden (1st ed.), 1820, (2nd ed.), 1929.

218. That is, investor evaluation of the comparative value of the purpose for which the funds are sought and the interest on the bonds, and competition in the capital market.

219. Alfred E. Smith, 1873–1944, governor of New York State and presidential candidate.

220. Jay Cooke, 1821–1905, U.S. financier.

221. Likely Martin Barnaby Madden (1855–1928), elected to Congress from Illinois in 1904 (Republican). He served on the Committee on Appropriations and Budget in the 66th Congress and was the chairman in the 67th Congress. Both times, major reform in the budgetary process was enacted. He was referred to as a "Watch Dog of the Treasury". He published two opinion pieces in the Saturday Evening Post: "Tax Reduction and the Public Debt" (Oct. 17, 1925) and "The Budget to Date" (Nov. 7, 1925). (*Not* Frederick William Madden, 1879-, with Bureau of Valuation, Transit Commission, State of New York. A New York State Assemblyman, Tracy Madden, died in 1916.) A quote from Martin Madden, claiming the end to what is now called pork-barrel legislation, is found in Carolyn Webber and Aaron Wildavsky, *A History of Taxation and Expenditure in the Western World*, New York: Simon and Schuster, 1986, p. 415.

222. The Cour des Comptes (Comptes – counting or accounting) was established to oversee taxes and financial laws in France. It has its roots in the early 13th century as a system of public accounting was developed. Ministers were appointed by the head of state and they were responsible for: (a) making judgements and settling disputes over taxation, and (b) making quarterly reports. Before 1789, there were thirteen offices, the largest being in Paris. It was abolished in 1791 and eventually replaced by the Bureau of National Accounting.